RHINESTONE QUALITY

LINDA WILKINSON

ISBN 979-8-218-70545-9
First printing, June 2025

To my family, Megan and Heather, for their unconditional support, even though they've never understood what in the hell I do.

With gratitude to all of the IT people who have touched my life and made me think (or laugh) in the midst of all the madness. A special thank-you to all of the quality professionals out there, heroes all, fighting the good fight.

And last, but not least, to Dave Beiers, who gave me a Kindle Scribe and said "Chop, chop!".

TABLE OF CONTENTS

CHAPTER ONE
DO YOU WANT TO BE A COWBOY?

What, exactly, is Rhinestone Quality? Quality, overall, is everything that touches a customer and the rest of the outside world to inform them as to the worth of your organization and products. Anything that can influence their opinion of your company and their desire to do business with you is part of the quality landscape.

In IT, quality normally means the quality of your systems or applications. The lowest common denominator in terms of this type of quality is that your applications do what they're supposed to do and don't break. Viewing quality strictly from this viewpoint may cause you to ignore everything else that influences your customer and makes them want to do business with you. Or not.

Rhinestone policies, processes, and products are what you've got when you talk the talk, but don't walk the walk. Do you remember the concept of a Rhinestone Cowboy? He was someone who looked like a (fancy) cowboy. He sounded like a cowboy. He presented himself as a cowboy. He might have sung cowboy-like songs. But it was all smoke and mirrors. He had never actually done any of the work or lived the life of a cowboy. It was all an act to draw you in and to get you to buy what was being sold.

Rhinestones are shiny, attractive things. They have a lot of bling. But in essence, they are not real gems and except as pretty baubles, they are valueless and lose their luster and worth over time.

You see this as it applies to businesses every day. Their stated values do not align with how they actually operate. They "kind of" invest in certain areas, without giving those efforts the commitment and weight really needed to make a difference – let alone incorporate them into their culture.

So why does quality matter anyway? From an employee or customer perspective, the answers are obvious. But why does it matter to a company? For-profit organizations exist to make money and proliferate themselves. They do not have feelings. They are an 'it'. Everyone who works for it is there to help it make more money and grow.

Quality matters because it is a differentiator in a highly competitive and wired-in world. You have to sell your products or services to human beings. Human beings will work for your company. Human beings will judge your company, talk to each other via a plethora of social media platforms, and your sales and profitability can rise or fall accordingly. You can gain or lose valuable personnel and/or partnerships. In short, your quality as an organization ultimately matters to your bottom line.

It sounds relatively simple on the surface and it is simple – to a degree. Do and be what you say you are. If you're going to invest in something, commit to it and do it right. That said, reality is a harsh mistress. Some company executives wish they could do things the right way and the company can't afford it. Some don't know what the right way is and struggle. Almost all will try to do things the fastest and cheapest way possible. This will happen over and over until a catastrophic event forces the executive team to take it seriously or they realize how much time, money, and effort they've spent for little or no return. The sad part of the equation is that it can take a company years to get to the point where they know they need to change. Such a loss of time and revenue!

There is nothing intrinsically wrong about wanting things fast. Or cheap. There are times when those are the best answers. What is missing is the realization that fast and cheap won't always work. Potential company profit needs to be invested in those things that provide a relatively clear ROI (Return On Investment) or sometimes, not invested in at all. It's the half-assed, shiny rhinestone nothing solutions that bleed companies of money that could have been spent elsewhere that drag them down and need to be avoided. It costs money, there's little or no return, and everyone involved just gets frustrated. On occasion, there are answers that cost the company nothing but commitment and a determination to turn their wishes into realities. It's not easy and sometimes it's not cheap, but you can turn rhinestones into diamonds.

This book is about turning into a diamond. As a company, as a boss, and as an employee. We all spend too much time and money on stuff that doesn't matter, just so we can say we do it and look good. The problem is you may not look good for long.

No one wants to buy worthless rhinestones. No one wants to do business with people selling them useless rhinestones – often with assurances that they've invested in priceless gems. We're surrounded by pretty, shiny things. We're not a pack of crows, however, and even a crow would spit out some of the things we've swallowed, hook, line, and sinker.

I'm going to focus on IT, because IT is what I know, but many of the problems and ways to solve those problems are universal. If you think this is going to be a book that does nothing but rag on "the man", you'll be disappointed. I'm going to rag on everyone. I've worked in IT for a long time, going from a low-level peon (and I was peed on a lot) to a VP. I'd like to share what I've learned along the way. I hope it helps you with your own journey. The ideas here are meant to help your company succeed, your boss succeed, and you succeed.

Without your company's success, you and all your compadres have no paycheck. As hard as it is to accept, your success is intrinsically linked to the success of the company you work with and for. You weren't hired to do what you want to do, however you want to do it, whenever you feel like doing it. You were hired to help the company achieve its goals. Your success is also intrinsically linked to the success of your boss. If they don't do their job, you'll never get a raise, promotion, help, training, or have a clue as to direction. The good news is that the success of the company and your boss is intrinsically linked to you – their employee. Without employees, all a company has is a big bag of ...nothing.   They'll have ideas with no way to execute against them.

Rhinestone quality, distilled to its core, is what you get when what you represent is a pack of lies, careful and intentional stretching of the truth, and flat-out misrepresentation of what is true. It looks good. It sounds good. Ultimately, however, facades are fragile and you have to spend an inordinate amount of time and money protecting them. Time and money you could be spending on something else more important to your success. Not only that, if your facades crumble, damage control is going to be expensive. Possibly so expensive you'll be unable to ever recover your market share or reputation.

What is the answer? It's really simple – but also profound. It's all about truth. Transparency? Not necessarily. Tell the truth about what you can and just shut up about everything else instead of making up stories or pseudo-philosophies trying to justify something you can't really justify in a way that reflects well on you and your company. You will inevitably just make things worse. Transparency is actually a weasel-word from a corporate perspective. There is no way to run a business with total transparency. There are times it is actually illegal or a breach of contract to share certain information with everyone. Transparency also can be interpreted to mean different things to different people. If you really, really want to claim you value transparency for some reason, decide in advance what that means to you and publish it somewhere like the employee handbook.

Take a look at some of the craziness in the news right now. One ginormous company had a mass layoff and said it was cutting underperformers. Really? Did they hire that many underperformers? Did the company really think that made them sound competent? Did it ring true – with anyone? It was embarrassing and horrifying. They now have some pending lawsuits. It would have been infinitely better to say the company was tightening up and reorganizing to meet changing priorities and were reluctantly forced to lay off a number of their staff. End of story. That company, and every other company, owes nothing to the evening news or anyone else.

News agencies like to make things sensational. They normally have no problem doing that by feeding the general public's prejudices about big business. Many average people just assume big companies or rich people are inherently evil. That means they will be predisposed to suck up any stories about heinous businesses and/or rich people and believe those things to be true. It certainly doesn't help a company's case when a mass layoff, which is bad enough, is accompanied by the heartless dissing of the human beings being let go. It's simply cruel and unnecessary. There is absolutely no need to insult or belittle people during a process that is painful at best. Layoffs, by their very nature, mean "no fault of the employee". Otherwise, you fire people with cause. There are companies that lay people off every year. These layoffs often occur during the holidays. Nothing about the process is kind. There is no need to make yourself look even worse by adding insult to injury or by making up a lot of shit that just makes you look like an unfeeling bastard. That is not in the best interests of you or your company.

There were recently some statements made about another giant in the field that is flattening its middle management. Suddenly their CEO is all over the news being attributed (whether true or untrue) of giving statements that don't make a lot of sense, especially if you've ever worked for any company, anywhere, ever.   The worst part of the entire thing is it's a company I've actually talked to 2 or 3 times at various times during my career, and they're right. They genuinely do need to re-evaluate their middle management layers. It happens a lot with companies that explode in size really quickly. There comes a point when you need to re-evaluate and rein it all in, cutting those positions that no longer make sense.

To this company's credit, they were doing their best to re-deploy and combine groups rather than lay people off, but some will undoubtedly fall to the knife.

Regardless, other than making a public statement saying after years of explosive growth they were re-evaluating their organization and shifting things around to be more efficient, no one needs any other insights into their company's problems, setup, or purposes. Instead, we're hearing all kinds of philosophies as to why the changes are being made. One I read said "Good managers do more with less.". Wow. What a load of crap. Generally speaking, even superstar managers who are the next closest thing to God do less with less. Weasel words. What's worse, there's no need to explain your company's decisions to me or anyone else, unless you're engaging in some sort of illegal activity or something that will impact the general public. Just say what needs to be said and then stop.

If you are as kind, understanding, and helpful as possible to those impacted, you've done everything another person can do and will help the rest of your employees retain their belief in your company and you as a human being. You can sleep at night knowing you've done everything you could under the circumstances. I happen to really love this particular company, as they are unusually customer-centric for such a huge organization, so I find it painful to watch them dig a pit and fling themselves in.

No doubt these companies consider this damage control. Again, maintaining facades are expensive and not especially effective. Say what is absolutely necessary, and if it's possible for you, show some genuine regret for human beings that have been negatively impacted by what is essentially a business decision. The last thing you or your company need is the rest of the world regarding you as a pack of weasels.

These few examples are just examples of backlash due to corporate decisions making the news, but as ugly as they might be, they pale in comparison with production catastrophes. Consider the recent problems of a large aircraft manufacturer. They had a door fall off an aircraft in flight that was full of people. The bolts had not been tightened. Everyone in the QA space or who has been through an RCA (Root Cause Analysis) understood this quality failure right away. It was the "why" that was the question. Either the step to check the bolts was not on the list, or it was on the list and was not done. Those were the only choices. In the first case, it needed to be added to the list. In the second case, there are more serious process problems that need to be addressed. Later, employees leaked out that people were encouraged to take shortcuts in the interests of time. The company, and those on the plane, were extraordinarily lucky. There was no loss of life.

The decisions that led to skipping things in order to make a date should sound mighty familiar to IT shops. They take these kinds of risks all the time. Should they? It depends. If human life is involved, my own answer would be no. Anything involving human life should be risk-averse. In other cases, yes, there are times where errors can be annoying, but not significant enough to halt moves to production. The key to success every case is to know what is, and is not, important. If you don't know what is important, or you don't know your customer and what they value, you will make mistakes.

This leads to talking about software quality specifically. Software Quality Assurance is the Rodney Dangerfield of the IT world. They don't get no respect. Unfortunately, they don't get no budget, people, or tools either. It's a constant battle for funding. It is the Poster Child for Rhinestone Quality, although any group or organization can suffer in the same way. Everyone wants quality. Everyone says they have it. But do they? Let's look at some examples.

A company decides, after a series of Unfortunate Events, that they need to focus on quality. During their budget process, they put aside enough money to hire a manager, director, or

VP. And that's it. They tell themselves and each other that the person they hire will tell them what they need. In other words, they say they are going to invest in quality and when the rubber meets the road, they put aside no money to fund it. If you are the hire they make, when you come back and tell them (for example) you need 12 people and X tools, for a total of 1.5 million dollars, you can pretty much expect them to breathe into paper bags once you've left the room. In order to get funding after a budget cycle, you will have to take it from other budgets. Guess what? None of the owners of those budgets are going to want to give you squat. You'll be lucky if you can get 2 people.   During the next budget go-round, maybe you'll get a few more. This doesn't happen in other areas. If you're going to start up a software engineering group, for example, it rarely begins with no budget to hire staff or purchase tools.

You can tell right away if you're going to have these kinds of issues. When you interview for one of these roles, ask about the budget. If they say they were expecting you to tell them what is needed, don't be seduced by thinking you're going to analyze the situation, hold out your hand, and money will magically drop into it. It's Rhinestone Rhetoric. It sounds great – doesn't it?

You can also look out for JDs (Job Descriptions) that say you'll have to wear many hats. How fun, right? What that actually means is that you might be the highest paid Quality Analyst or SDET (Software Development Engineer in Test) of all time. It means they do not have enough staff, have not planned enough staff, and may never have enough staff to do the job(s) at hand. Furthermore, they are looking for you to provide tactical solutions, rather than working on the strategic initiatives that should be your job. In other words, people outside your group will be providing strategic direction for your group when they don't necessarily understand or know how to support the work. This is not some kind of rare situation. This happens all the time. It is common.

Let's make this even worse.

QA is one of the few fields where anyone who has even tested your patience thinks they know everything there is to know about your job. They are going to tell you what you should be doing and how you should be doing it. They are going to be telling your people how to do their jobs. They are often going to ignore what members of their quality organization have to tell them. They are going to pressure you to say or do things that are wrong in order to make dates that are often artificial, and even if they aren't, don't matter in the face of the quality issues that have been uncovered.

In other words, you have to be tough. You have to be willing to educate everyone about what you do and how and why you do it for your entire career. If you're the type of person that caves under pressure or are submissive, QA is not for you. QA staff are not paid to tell people what they want to hear. They are paid to tell the truth. In some ways, that makes QA the best career there is – how many people in the world are paid to tell the truth? In other ways, it's the worst career there is – you'll likely never be regarded as a team player, many people aren't going to appreciate your truths, and you'll need to deal with a bunch of arrogant nits who know nothing about your work telling you how to do it. On the bright side, over time you'll be paid very well to have your expertise ignored. On the dark side, we all undoubtedly should have invested in therapy for our masochistic tendencies. Since all of the QA people reading this are crying into their adult beverages right now and are convinced I'm the Queen of QA, let's move on.

We've considered companies as a whole, quality assurance as a part of that whole, and that now brings us to management. For the purposes of this discussion, management means anyone who has human beings reporting to them. Many of whom are wearing rhinestone-encrusted cowboy boots. Just sayin'.

There are managers that are great, and there are managers that are Evil in Human Form. Unfortunately, the latter outnumber the former. Let's look at why and potential solutions for those problems.

If a manager (remember, the title doesn't matter, this can be a COO or a VP too) is part of a core team, they sacrificed time, money, and talent to make the company what it has become. Their position likely started as a buddy to the founder. There is nothing wrong with that. If you were starting a company, you would likely do exactly the same thing. Call some people you know and trust who have the right skills to get things off the ground.

These people are untouchables unless they actually break the law or become such an obvious liability they absolutely have to be let go, which rarely happens. These core people might have had very technical, specific skills that brought the company out of the idea stage to the reality stage. Their current title or responsibility is often a testament to their efforts and loyalty.

What you need to remember about these core members is they may have never had the customary training and mentoring that goes into actually interacting with and managing direct reports daily. What they are used to is handling nitty-gritty details and having total control over those details.

These types of managers tend to be micro-managers. They've hired you to do X job, but they don't really trust you to do X job. The only person they trust to do X job is themselves. In a more normal progression (starting at peon growing to demi-god), they would have learned to let go. It is one of the hardest things to learn when you move from an individual contributor role to a management role.

You might have been the hottest thing going technically and those taking your place, doing your former job, might not be as good at it as you were. That's OK. If you were really brilliant at your job, you need to recognize that most people are going to be normal people with good, not stellar, skills. Those of us that moved up the ladder gradually had to learn this early. You cannot do someone's job for them. You cannot do everyone's job and do them well. Something will suffer – usually your own job. You should be working on strategic initiatives and may not know how.

Micro-managers are common in many fields, but are especially common in IT, which is a detail-oriented field, period. These kinds of managers are more comfortable being one of the guys, shooting the technical shit, and tend to react very badly to things not going their way, as they're unaccustomed to dealing with other people.

For example, they might check your code and brag about writing code themselves. In reality, they should be ashamed of those behaviors. First, when they hire someone to do a job, they need to leave them alone to do their job. It's highly unlikely they need any help and if they do, it should be offered by a senior or a lead that is non-threatening and being paid to ensure work products are done correctly and on time. Not you.

What's more, upper managers need to train themselves not to undermine the authority of those that report to them. If you hire someone with 20 years of experience, for example, they aren't going to need your help to either build or manage their group. That might make you crazy, but if you made the right hire, they are not going to need you to run their own group. You do not need to have 1/1s with their people. You should not be reviewing the individual work products of their people in your spare time (of which you should have none) and critiquing them. You don't need to interview their potential hires.

Those are immediate and obvious signs that you do not trust your direct report, let alone their people. You have just insulted and made that person paranoid. You will likely terrify their staff. Do you really have so little to do that you feel you need to do someone else's job?

Throw that rhinestone cowboy hat in the nearest recycle bin and go do your own job. If you've done things right and made the right hires, no one needs you to do theirs. If you can't break yourself of your micro-manager tendencies, sit back and decide what kind of information and metrics would make you feel better, and ask your direct reports for that information. It's worth noting that micro-managers are not just disliked, they are generally hated by their staff. Back off and let your people breathe.

Which leads us to managers that need to feel superior to their people. It's a power thing. These are the managers that need to make you feel bad in order to feel good. Their primary skill set appears to be belittling others. Usually, these folks mistake being obeyed for being respected. They rarely understand the work underway and compensate for their insecurity by "bossing" others around. They are bullies.

Companies that really want managers need to invest in managers. If you are promoting your core team into management spots when they have no management experience, get them trained, or get them partnered with an experienced, seasoned, respected manager who can help mentor them. Afterwards, pay attention and see if they're successful working with human beings.

If you're making a management hire, your best bet is to hire someone that likes and cares about other people. Ask for examples of their mentoring, promotion, and management of staff. How they build a killer team. How they keep them motivated. How they handle personnel issues and problems. Managers must be able to manage people. Otherwise, they might have the title, but they can't do the job. Rhinestone management.

Finally, while we're talking about rhinestone quality, we need to talk about employees and what they might want to remember in terms of what rhinestone quality actually means. They should learn it before they go out on social media, telling anyone who will listen about how heinous things are and all about the internal operations of the company in which they work.

As an employee, it's unethical, to say the least, to take someone's money and time on the one hand, and try to hurt them on the other. That pretty much makes you a candidate for Worst. Employee. Ever. If your company is genuinely that bad, get the hell out of there. Leave a warning on Glassdoor on your way out to save innocents from making the same mistakes you did. Don't bite the hand that feeds you – at least wait until you're no longer taking their money to try and do them some sort of injury. Show some class and some integrity. Rhinestone quality involves everything that touches the customer. That includes you.

You may be a pretty, shiny, talented person. Maybe you haven't been treated well. This is your opportunity to rise above those that were evil to you and refrain from becoming evil yourself. The person you'll hurt the most is yourself. No company will miss such an employee – they'll feel justified and like they've dodged a bullet. You will be viewed as a detriment to the company – and any injury, if any, you actually inflict upon them you will also be inflicting on everyone who works there, including people that presumably were your friends.

And geesh, get a clue and do your postings and pontifications on your own time and not while you're on the "company clock". Your assumption your boss or work buddies will never see or know about your media activities is about nil. If you're remote, investing time, money, or effort in finding ways to prove you're online working when you aren't is likely to backfire on you in a big way – deservedly so. It makes you a liar and a cheat. If you work in an awful place, becoming an unreliable, treacherous pig yourself is not the answer. If you rock at your job, the best revenge you have is when the company loses you and your talent.

If you stoop to lying and cheating, you're no loss at all. Don't be that person. Don't become a fake, rhinestone employee. Don't applaud for people who are. The only person you hurt is yourself, so if it's revenge you want, remember "Revenge is a

[15]

dish best served cold.". Plot your escape and then execute it. Move on. Be happy. Happiness and success are the best revenge – and you are the one who wins.

Now that we've opened the Pandora's box and spread the fertilizer around to virtually every aspect of business, let's take a look at some specifics. Read on...

CHAPTER TWO
THIS WON'T HURT A BIT...

Do you read a lot? Keep up with field? If so, you likely read on LinkedIn, Glassdoor, and, if you're like me, a variety of IT-related sites and blogs. Does it depress you? A lot of it just makes me sad, to tell you the truth.

Business has not changed, or if it is changing, it is at a microscopic pace that is difficult to discern. The technology doesn't matter. What technology has done for us is to allow an outlet for anyone and everyone to put their opinions and feelings, no matter how misguided or immature, out there for everyone to read and respond to. In this regard, technology hasn't done us any favors.

It occurred to me that with the advent of the pandemic and resulting WFH (Work from Home) mindset, many employees are not observing or learning corporate culture or getting the kind of mentoring they need to survive. So I thought I would spend a little time talking about business in general. Company people ("the company") are not going to like what I have to say and people that work for a company are going to like it even less. At least I resisted the temptation to name this chapter "Bend Over and Grease Up". Overall, I get the feel that the lack of understanding of what a company is and how it survives is so prevalent it is worth a chapter to talk about it and possibly help some people understand and have better success and less angst working within the confines of someone else's company.

I'm going to focus on IT here, because that is my niche and it provides a model   easy to understand. Someone has an idea.

They may or may not have the skills to produce that idea. Often, they will start by calling a few buddies, selling them on the idea, and they may work on the idea in their spare time. When they have some sort of product, they may offer it for free to someone who could perhaps benefit, asking for their recommendation to others in return. If their product has some success, they add people as necessary to support their growth.

They pick people to help them that they know have the skills they need at that time, and that they trust and like. Everyone focuses on making the vision a reality. These core people form the nucleus of the new company and have often sacrificed money, time, and a whole lot more to make it happen. When they start adding people, it will always begin with others this core group knows, trusts, and likes. What you have here is the start of an "old boy network". The Old Boy Network is alive and thriving at this time and it has nothing whatsoever to do with age. Young companies have it and old, established companies have it. The age of the "boy" is immaterial.

I read tons of crap about baby boomers, and most of it shows a lack of understanding of what business is and how it operates. Do not be that ageist, prejudiced jerk and show the world you aren't an immature newbie. Pay attention to what is going on. Problems you're experiencing have been problems for many generations, including your parents, grandparents, and great-grandparents. They will likely be problems for your children and grandchildren, unless you and whatever generation you represent change the world. Um, good luck storming the castle.

Businesses start with some core people who know, like, and trust each other. They start their growth by adding people they know, like, and trust. At a certain point, they will need skills their network cannot provide, and start hiring some people they do not know or particularly trust, but they will try to choose people they like during the hiring process. These people may or may not become part of the core network, depending on the level of dedication, commitment, and skill they bring to the table and whether the core team can like and trust them.

Before you critique this, consider what you would do if in the same position. It is likely the same thing; you'd reach out to people you know and like and who have the skills you need at that time. The ugly truth, however, is that beyond that Old Boy Network, no particular job is necessarily secure. The higher up the ladder you are, the less secure the footing. There is a reason C-level executives, VPs, etc. make the big bucks. Their jobs are inherently fraught with risk and that can result in a lot of paranoia.

So the company grows and eventually becomes an "it". A company is not a person. It does not have feelings, it doesn't have empathy, and it exists to make money, grow, and protect itself. Consider, however, that if it does not do those things successfully, you are likely out a job. A company owes its stakeholders first and everyone else after.

In terms of a company, you, as an individual, do not matter. That does not mean no one cares about you. It means that if times get tough and losing you or your job function benefits the company, that is what is going to happen. Many companies lay people off every year. Often right around the holidays. Does that seem humane, decent, or caring? It isn't. And it has nothing whatsoever to do with your skills or value. It has to do with what is in the best interests of the company at that point in time.

Often a company needs to lay people off in order to show a certain percentage of profit or pay out certain dividends the end of a year. If they don't do that, they may lose beneficial interest rates, customers, investors, or the like. It is not personal, even though it feels mighty personal. What is even more horrible is that they are going to have to find some way to explain your departure so that the company looks strong.

This is particularly true with a mass layoff, but it will be given some consideration regardless of who or how many are let go. The company has to explain your departure in a way that does not terrorize the remainder of the staff and result in resignations they didn't want. They need to look strong and healthy to their stockholders and the rest of the world. This means they are going to give reasons for your departure that might bear absolutely no resemblance to actual truth. Again, this is a strategic marketing ploy and is not personal, no matter how it actually impacts you, the human being(s) involved. The company is an "it". You shouldn't expect feelings or understanding from an inanimate object.

Your boss may genuinely care about you and may have been spending their time trying everything they know to save you and your job. Your peers might cry when you hit the door. Ultimately, however, whatever benefits the company will prevail.

In short, it simply isn't about you. You need to understand this basic truth. That doesn't mean a company will not make some effort to keep employees happy. The people who run and manage companies aren't stupid and they know happy employees do better work. If they can reach their goals AND make you happy, they will. But if not, corporate goals are more important and it's vital you understand part of the reason your boss (and you!) were hired was to further corporate goals. Not your own.

This is not all bad news for the employee, just a shift in understanding. You and your position are an exchange of money for skills. You, and only you, are the captain of your own ship. While you will be provided with money and benefits, your actual happiness, career growth, and health are not the responsibility of the company. All it offers is money and benefits in return for your skills. A business transaction. If you enjoy the work and the people, that's fantastic and it's all gravy. If you are not happy with your work or the company, it is up to you to change that and it means you might have to leave.

Unless the company is breaking the law, it is not required to change the way it operates (or wants to operate) in order to make you, one individual, happy. Find work that makes you happy. If you are not making enough money or do not have enough benefits, it is up to you to talk to the company and see if that can be changed if you want to stay, or to leave for greener pastures. All of that is on you. If you feel you deserve to be promoted and aren't – talk to someone first and see what you need to do and if you're unhappy with the results, go out and find a better opportunity.

It is infinitely easier to do nothing and complain than it is to go out and make things happen. It doesn't help you and doesn't help your company if you just stick around and whine. Become a dynamic mover and shaker. Move on if you are unhappy. Wishful thinking is just that – wishful. Make your dreams a reality. It is not realistic to think your situation is going to magically change. It won't.

If you want to increase your potential happiness, look for companies that align with your value system. If all you care about at this juncture is money (and all of us have been in that boat at some time during our careers), that is the easiest kind of new opportunity to find. There's always someone out there willing to pay you more money. Be aware, however, that money, in and of itself, will not make you happy. I've had jobs where no amount of money (and I'm not joking here) would entice me to sign back up.

If the mission and purpose of the company matters to you, look for a company you, personally, would be proud to represent. While companies in general are an "it", some companies have a vibe that can make you happy, provide a service or product that means something to you, or are being run by people who give a damn about their employees. These opportunities are harder to find, but if you know what makes you happy and you concentrate on companies that can meet your standards while you meet theirs, great things can happen.

While I haven't had any illusions about companies, which are inanimate objects, for quite some time, I have reported to people I genuinely liked and respected and had the privilege of working with some really amazing people. I've worked with several companies that had a "vibe" that resonated with me and made me happy. I've made a lot of money. I've received many awards. When I look at my career, I feel good about it. You can achieve that too, but not unless you take charge of your own career.

When you get kicked in the head, pick yourself up, learn what you can from the experience, and move on, trying to improve your situation in some way each time. There are times you may need to make a calculated move in order to achieve something specific in your career. Do it. It's all on you.

Something that many people don't understand is that while you collect a paycheck from someone, a measure of investment and loyalty is appreciated and often rewarded by a company. It isn't your company. They hired you because they were hoping you would be able to help them achieve certain goals. Not your goals – their goals. If you are not committed to doing your best work and whatever you can to make the company successful, no doubt someone else out there that could use the work would be willing to help them move forward.

If it were your company, would you hire someone who just wanted to do what they wanted to do and didn't care about the company? I wouldn't. Would you hire someone who dogs you and your company on every social media platform they can find? I wouldn't. So don't be immature about what the company "owes" you. If it pays you, provides a safe place to work, and gives you benefits, they're pretty much meeting their obligations. Everything else is up to you and your choices, good and bad.

Take charge of your own life and do what you need to do to make your life worth living.

While we're talking about the company and acknowledging that it is an "it" and not a person, let's delve a little deeper and talk a few specifics. One of the most common misconceptions in business is the function and purpose of HR. Human Resources, which may have a more politically-attractive name like People and Culture, exists primarily to protect the company. Not you as an individual. If you have problems with your W2 or benefits or need to put through a name change or something similar, HR is going to help you. If you are having problems with a co-worker or your boss, they tend to support from the top down unless there is a legal reason for them to support you. While it is easy to play the blame game here, remember that is part of their job and the reason they were hired.

When you are having a problem with a co-worker, whether it is harassment or something else, in order to protect yourself, start keeping a log. Your first step is to let that co-worker know their behavior or actions are unacceptable to you. No one has ESP and they may not know they are making you uncomfortable. Give them a chance to change first.

If that doesn't work and the person is NOT your boss, go to your boss and bring your log. Why is this important? If you have not told the individual irritating you to Cease and Desist, the first thing your boss is going to ask is if you have talked to that individual. If you say no, a decent manager is going to encourage you to do that. Otherwise, you're going to be stepped upon during your entire career.

Confrontation is difficult; there are entire classes out there that deal with handling conflict specifically. It is a critical "soft skill" for moving up the ladder. That said, if you're too intimidated or uncomfortable with speaking to the individual yourself, your boss should step in and help you. Removing impediments to getting work done and ensuring a safe environment is part of their job.

If you skip your boss and go directly to HR, the first thing HR is likely to ask is if you have spoken to the individual and your boss. If you say no, a decent HR department is going to encourage you to do that for the same reasons. If they don't, regardless of what they might say to you, their first action is likely going to be to talk to your boss. If you've skipped your boss and gone directly to HR, your boss will be embarrassed by this. It gives the impression you two don't speak and you don't trust them to do their job. No boss is going to appreciate that, even if they understand it. What I'm saying here is that if you have a hierarchy, follow it. If you do not follow the hierarchy, expect some fallout from your lapse of judgement.

Now consider a problem that does involve your boss. One step applies regardless – you must tell an individual that their behavior or actions are making you uncomfortable and ask them to stop. Give serious consideration before you take any action as to whether the incident(s) are actually illegal or if they just annoy you. The only way an HR department is actually going to get involved and help you is that if NOT helping you could result in a lawsuit or some sort of legal action or sanction. Otherwise, your boss is going to prevail. There is no point in fighting with your boss. While not always true, generally you will lose those battles every time.

Once you engage in this type of battle, your tenure in that company is likely to come to an ugly end. Yes, there are rules regarding not punishing those that raise problems, but that's Rhinestone Policy and wishful thinking on your part. Remember that if a company wants to get rid of you, they are going to find a way. If you are getting in their way, they are going to move you out of their way. It's horrible, it's not fair, and oh yes, it's true. This means you really need to think about what your desired outcome is for taking action.

If your boss is doing something illegal, then it is realistic (if you've been keeping a log and have your ducks in a row) to expect the company to take action. They will do this to protect the company, not necessarily to protect you. They will also remove them if they really wanted to get rid of them anyway. They'll use your testimony to get that done. That said, they will not particularly want to keep you or have you reporting to them either. If you tanked one of their peers, they likely will not trust you won't do the same to others. It is extremely tough to be a whistle-blower and you need to be prepared for the fallout.

Now that I've said all kinds of horrible things about HR, I'm going to totally switch gears and look at things from another direction. Heading up HR is one of the worst, most stressful jobs in the company, particularly if you really care about people. If a surgeon is working on you, you don't want them to be so empathetic they cry over your incisions. People who work in stressful situations develop weird humor and ways to compensate for what they need to do every day. Those of you that work in clinical situations know what I mean. HR personnel are PEOPLE. They have feelings, just like you do. Sometimes the things they have to do to protect the company, like head up mass layoffs, keep them awake at night for months. They need to develop some sort of detachment in order to survive.

That said, I have known several amazing, remarkable HR professionals that have handled the hard stuff with grace and caring (without breaking down or becoming unprofessional) AND been a helpful, empathetic employee advocate at the same time. I've been in meetings where hard decisions were being made and heard them speak up for the employees. I've seen them knock themselves out trying to find the best benefits possible on limited budgets. I've watched them advocate to get someone something they deserved, or speak to someone about a problem without damaging their career or future. But wow, those kinds of people are rare. If you have one, lucky you. Otherwise, it is always best to fly under the radar of HR.

Try to handle things yourself first, then involve your boss, and use HR as a last resort. It will make you a stronger person anyway and if you're interested in a career as opposed to just a job, the ability to handle conflict positions you for growth. Unlike what many view as management – sitting around with your feet up whilst your eager minions do your bidding - management is actually handling problems every day, all day long. By the way, if you ever find eager minions to do your bidding, I'd love to hear about it. QA people make notoriously crappy minions.

So embrace (no matter how difficult) and confront your problems and learn how to solve them with confidence and grace. Again, this is an extremely valuable skill that positions you for bigger and better things.

Now that we've dealt with the company and addressed HR, let's talk about people. And politics. Even a company that doesn't have much in the way of politics has politics. And some companies have politics akin to shark feeding frenzies.

My best advice is to be observant, to be smart, and to pick your battles. There are going to be "untouchables". These are people that were with the company before you came on board and are going to be there once you're gone. Do your best not to poke at them. I think it is also helpful to genuinely consider each situation from a corporate (not personal) perspective and support whatever is in the best interests of the company. No matter how much that Spawn of Satan wants to get his own way, preferably over your dead, bleeding body, if you genuinely have the best interests of the company (as opposed to your own interests) at the forefront, it will be respected and you may emerge with all your body parts intact.

You need to pick your allies with care and remember that if things come down to you or them, they will choose themselves. Again, human beings are capable of remorse and they may not like it, but in some companies, survival instincts run strong. Do not be surprised if sometime during your career you get stabbed in the back by someone you liked and trusted. This is commonplace. It hurts. Learn from it, don't carry grudges (it hurts you more than the person in question), and move on a sadder and wiser person.

Drilling down a little deeper, you, as an individual, have choices to make. Do you want a job? Or do you want a career? No one is going to hand you a career unless you have a family member that owns the company. You're going to have to earn a career.

As I read all of the social and professional media out there, I realize many people don't have a clue. No one is going to hand you anything. It took me years of hard work and determination to get to the status level and pay I've enjoyed. Years. There are no shortcuts that don't involve family or something kinda disgusting. During that time, you're going to have to work really hard, learn a whole lot, and deal with some totally poisonous assholes. Lots of things that happen to you will be unfair, horrible, and borderline revolting. If you're part of a minority, like a woman in IT, you're in for Quite the Time.

Overall, remember you are the captain of your own ship. You do not have to stay anywhere – you are not a slave. If you cannot do without a job, look for another one while still employed until you find something with more promise than what you've got. Often the only way to get ahead is to move. And remember, it's all on you. The company owes you a job – not a career. That's in your own hands.

If home/work balance is your primary goal and you do not care about a "career", then go for it. You are likely to be happy and successful if you are not ambitious and making a decent but not impressive salary is enough. I know a number of people who fall into this category and I respect it. A lot. Most people are not that self-aware.

If you ARE ambitious, however, there is really no such thing as work/life balance – especially in IT. You're going to have to pour effort and sacrifice into a career. You're going to have to keep yourself educated and on top of technology if you're in IT, and a lot of that will involve your own personal time and investment in yourself.
We've already pointed out it is a myth that managers (and up) sit around with their feet up whilst their minions do all the work. They are on call 24X7, will likely laugh if you ask them about their own work/life balance, and they are responsible for everything – good and bad – that happens within their purview.

So if you screw something up, it's highly likely your boss is explaining that to someone whether you know about it or not. A good boss protects and advocates for their people. You know all those "worthless" meetings your managers have? They might be fighting to get you a promotion. Raises for their people. More people to help out. Figuring out with their peers what your next projects are going to be. What to do about a problem – internal or external. So much of what they do is going to impact your life and that of your peers. They are, in a way, the least useful members of the staff. When you get a good one, however, they are also the glue. A bad one will be more concerned about themselves than they are with anyone else or the company.

You are going to need to remember that not every manager is good with people. It should be a requirement, but it isn't. Some were hired to provide a skill needed at the time and were known and trusted or are part of the "core". That doesn't make them good managers. I've worked with and for managers that shouldn't be allowed anywhere near other human beings. It's just part of reality. So if you have a decent manager, be thankful and appreciative. It isn't all that common, given how companies start and grow. If you work for a bad one, get yourself the hell out of there. They are highly unlikely to suddenly have an epiphany and change their evil ways.

We've talked about generalities here, and since this book is about QA specifically, I'd like to finish up by talking about QA (Quality Assurance). QA is kind of an odd duck and managing a QA area and QA people is not the same as managing other areas.

We've talked about the company, the "core" team, and politics. Now let's explore QA specifically.
Generally speaking, QA people are constitutionally incapable of dealing with politics. From your lowest-level tester to your highest-level VP, they are unable to survive typical corporate politics and will require protection in order to do their jobs and thrive. A good chunk of your time as a manager of QA folks will be preventing them from shooting themselves in the foot and getting themselves fired.

I attended a conference talk a year ago where the speaker said that if you work in QA (regardless of position), you have to be prepared to be let go when you become inconvenient. That is a genuinely terrible thing to say, and the reason it is terrible is because it is true. Fairness doesn't really come into play here. If a company wants to dispense with your services for any reason whatsoever, they will find a way to make that happen. This is corporate reality and you can rage against the machine, or you can accept the reality and do what you need to do to keep yourself relatively sane and moving forward.

Working for any company, unless you are part of the core team, is not really fair or fun. Your coworkers, bosses, environment, and the corporate vision can be fun. But the company is an "it". And it does not exist to make you happy.

QA is poorly understood. After all these years, it is still poorly understood. Everyone who has ever tested anything thinks they know everything about it and are therefore qualified to tell you how to do your job. Its demeaning, insulting, and ignorant. There's just no nice way to say it, so "Suck it up, buttercup.". A good 75% of your job will be attempting to educate people who do not want to be educated.

So here is lesson number one for QA. You have to be tough if your aim is to make it your career. You have to be patient. And you have to put up with a lot of crap. If you manage QA people and have never actually had a job in QA, think about what life is like for a QA person and try to walk a mile in their shoes.

The job of a typical QA Analyst or tester is to detect anomalies and report them. They have been trained to tell you when your baby is ugly, whether you like it or not and regardless of your position. To top things off, they are accustomed to being lied to – everyone lies to QA. Developers tell them they haven't touched (whatever) or that their code is golden all the time. Management staff lie to them and pressure them to skip tests or ignore the "unimportant" things all the time. This makes them skeptical by nature.

They will inevitably ask inconvenient questions at the worst possible times in an effort to figure out whatever doesn't make sense. That will include the polite fictions put out by corporate to cover truth that cannot be made public. Overall, it's just a bad idea to lie to QA staff. Other groups will hear what they want to hear or whatever is most comfortable for them to believe, but not QA. They will understand if you can't discuss something for legal reasons, but have problems and lots of questions for anything that doesn't add up.

Consider sending out an email with a typo. You won't get just one person helpfully pointing it out, your entire QA staff will send messages to point it out. They can't help themselves.

On the plus side, QA people have something that you cannot buy with a mere paycheck. Integrity.   They will always, regardless of the risk to themselves, tell the truth as they see it and will attempt to do the right thing. Are they always right? No. No human being is always right. But they are right more often than not. These are people genuinely committed to what they do and what they do is different than everyone else in your company.

On an agile team, for example, you have developers building something. A Product Owner dedicated to building something. Possibly Marketing Associates building marketing strategies for the finished product. All builders of some kind, focused on getting a product out the door. What your QA staff do is not build and promote. They prevent and protect. The most important thing they do is PROHIBIT something from going to Production as is, due to problems that would affect your customers, the company, and ultimately your bottom line.

This is one of the reasons I'm not especially fond of decentralized QA staff. It can be mighty lonely to the be the only destroyer on a team. QA staff need others to collaborate with that have and understand their mission. They need to report to someone who will help them advocate for important issues, who understands their work, and who promotes and makes sure they are recognized for outstanding work.

Whether you work in QA or you work in another area of the company, it is especially difficult to manage or help anyone with integrity in a company that has none. Companies, as entities, lack any sort of human emotion whatsoever. They don't have consciences. That means the vibe or soul of a company or a department is largely going to be up to the people who populate it.

This means that when you accept a job, make sure you're getting a good vibe from the people you will be working with and/or for. They may turn out to be piss-poor excuses for human beings later on, but at least you've done all you can to make good choices. People tend to forget that the interview process should work both ways. They want to get a feel for what kind of an employee you'll be, and you should be getting a feel for what kind of an employer they'll be. You should be energized, excited, and feeling strongly like you "clicked" with your interviewers. If you just take anything ("any port in a storm"), the results may not be as promising as you might hope.

Ultimately, all we can do is the best we can with what we've got. Just be smart, be realistic, and remember you can choose to be a force for good in this world and make a difference to everyone you touch. It's up to you whether that is as a positive or negative role model for others. Companies, in and of themselves, are incapable of good deeds. It takes the people working within them to make a difference.

CHAPTER THREE
IT'S NOT ABOUT YOU

The idea of working from home (WFH) is a perfect example of a Rhinestone Policy. In both directions. Advocates say it's the hottest thing since sliced bread. Detractors say it is the root of all evil. In reality, it's neither good nor bad in and of itself. Success is dependent on the company, what needs to be done, and the individuals themselves.

There have always been some companies with a WFH (Work From Home) model. That might include off-shoring, small companies that are located in areas that are light in talents they require, and the like. The big push for WFH came around as a response to COVID 19, a world-wide, highly contagious, and deadly disease. Companies were forced into Work from Home strategies because the choices were WFH or die.

WFH was not the choice of the majority of companies. That said, business exists to make more money and to grow. If the WFH model was enabling that, everyone would be all over it. It isn't. It's really that simple. No amount of blog posts, complaints, threats, or reports really matter. Financials and the bottom line are what matters from a corporate perspective.

So business, which exists to profit and proliferate itself, is making moves to return to something which, for them, was a more successful model.

This is going to be hard for some to comprehend and accept. You, as an individual, do not matter. What you do and do not like doesn't matter unless you own the workplace. Individuals within your company can and likely do care about you. They may even agree with you privately. Ultimately, however, what is best for the company is the direction in which it needs to move. With or without you. The individuals in your company may care enough to hope it will be with you, but ultimately the company will make decisions in the best interest of the company.

The huge number of posts I've read on this — everywhere — are puerile, self-centered, and frankly, just ignorant. You are totally missing the point if your post is all about you, you, you. It's not about you. Those posts make anyone who understands and has worked for companies for a while assume you are young, self-centered, and not especially bright. Is that what you want?

Decisions of this type are just not about you. They are about the future and health of the company. Their future and health are important to everyone that collects a paycheck and likely the community in which the company does business.

Let's look at what happens when a company decides to open a business. When a company decides to locate in a certain region, the tax incentives offered to do so are often a factor in deciding where to go. States and regions sometimes compete against each other to offer a company the most attractive incentives. Why do you think that is? Because it brings people into the area — to live, to work, and to support other businesses. These incentives can equate to millions of dollars to a company. It can also help a region grow and thrive. Win-win. The company often gives back to community as well, making the entire transaction even more appealing.

If everything changes to WFH, that model crumbles. Regions won't grow, small supportive businesses die, and the community loses its support. How do you think they feel about your blog comment about how you don't want a 40-minute commute to work? Yup. Like you're a selfish, whining plebe that doesn't care about the company, the region, other businesses, or anything other than themselves. I'm just being honest here. None of us enjoy commuting. Your boss doesn't like it either. Neither does their boss.

We are all exchanging our skill sets and time in exchange for a paycheck and when we sign up with a company, we are also signing up to follow their rules. Not our own. If the pay, conditions, or benefits do not suit your lifestyle and are unacceptable to you, you are free to look for something more in keeping with your aesthetic.

Making the assumption that not supporting WFH is "old-fashioned" or something generational is just ageist, silly crap. Get over it; don't be the kind of person that blames everything they don't like on some generation that came before (or after) them. To support or not support WFH is a business decision and it has to do with money, reputation, and regional/tax support. I don't think it's a great idea to ignore all of the information these companies have about the success (or lack thereof) WFH had on their business. Companies always move in directions that benefit them financially. Always. You can assume if they are pulling back on WFH, it financially benefits them and their community in some way.

Let's (finally) talk about you. Personally, I vastly prefer work from home. I totally get it. But I'm firmly in the in-office camp. After years of working under both conditions, I've reluctantly come to the conclusion that most (that's right, MOST) people lack the discipline to do good work from home. I can actually hear the outraged screams from here. That doesn't apply to you, right? Yes, it does. I'm not buying what you're selling.

Work/Life balance? Do you mean the company should pay you to wash your clothes, prepare dinner, watch TV, take the dog for a walk, or bring up your children? Um, not. I've known one – that's right, one – person who actually committed to and gave a company 8 hours or more of solid work per day from home. So I'm calling bullshit on that "I get much more work done from home.". No, you probably don't. As hard as it is to believe, I say that with understanding and affection.

You want to work from home because you don't especially want to dress, shower, commute, interact with people and work 8+ hour days in one straight block of time. If you actually commit to at least 8 hours per day and you work from home, your days undoubtedly stretch out pretty far into the evening. It's really difficult to unplug when your living situation is also your working situation and you're fitting your work into your home activities. There is no distance between church and state.

Most people, however, do not commit to 8 hours per day. They do as much as they absolutely have to do and no more. It's almost like receiving a full-time check for a part-time commitment. Who wouldn't want that?

I've read posts that say as long as the work is done, who cares? What, exactly, does that mean? If you're in the office and you finish your work in an hour, you spend the other 7 helping your compadres, learning something new, or getting ahead on the next thing. It doesn't mean you pack it in for the day and go home.

Even when you're salaried, the expectation is generally that you will give the company around 40 hours a week of your skill and effort. If you're in IT, that's likely to be more like 60. If all that gets done is a very specific work task, no one ever expands their knowledge base, collaborates on anything new, or has an opportunity to learn and grow.

Unfortunately, since Ageism is alive and well, I'm sure many of you assume I feel that way because I'm an old bag who can't change. Not so, Grasshopper. The ability and willingness to change has nothing to do with age. It has more to do with experience and security. If you only know one way to do something, you are generally change-averse. It's hard to move from something familiar to something unfamiliar; it makes you feel insecure. That's true whether you're 24 or 64.

Almost every person I know, regardless of age, prefers to work from home. That does not mean, however, it is in the best interests of the company. It means it makes human beings happy.

Unfortunately, companies exist to make money, grow, and protect themselves. They do not exist to make you happy, except, maybe, as a customer. If WFH does not result in more getting done, more innovation, more money for the company, or more benefit for the region, it is a bad model for that particular company.

As a manager or a team member, one of the things about working from home that drags down the team and slows them up is the inability to reach someone who *should* be available. The team needs an answer to something and they get (crickets) for hours or an answer the next day. It negatively impacts the entire team. This happens with highly distributed teams as well. It's the polar opposite of "agile". Over time, I've become convinced that teams need to be co-located and at the very least in the same time zone. That doesn't mean you can't hire someone in some other zone. It means you should ideally build an entire team in that area. Take a look at the original concepts behind "agile". They were based on tight, co-located teams.

Another negative aspect of the WFH mindset has to do with finding a new job. Those who apply to remote-first positions are competing against a global market. Many applicants will make less than you and it takes a company quite a while to narrow down candidates to a manageable number of qualified people. Ironically, many of the people who complain bitterly about not being able to work from home also complain about how difficult it is to find work. If you can WFH, your competition has grown exponentially.

If you want a better chance of being considered, apply for local on-site opportunities where your location gives you an advantage or make it clear in your cover letter you'd relocate for an opportunity to work with their company. Relocating for an awesome job opportunity is common. It's not fun and it can be tough on you and your family, but it can be lucrative. Assuming you need to work, expanding your horizons and looking for work outside of what is "ideal" in terms of location can ultimately benefit you. You need to be strategic about your job search. Start with looking into local opportunities. Then open it up to WFH opportunities. If neither of those are successful, sit down and think about where you would (and would not) consider moving and expand your search. Those opportunities that are local, however, will always have less competition and the applicants will be in the same ballpark as you are in terms of salary. It evens the playing field.

If you restrict your search to WFH opportunities, you get what you get; a ton of competition from all over the world with all kinds of salary expectations, many of which may be modest in comparison with the area in which you live.

While we're talking about WFH and the current madness in terms of job availability, stop blaming HR for the crazy job market. They follow company processes and WFH fallout (to say nothing of thousands of applicants) are not their fault. It's hard to get back to 3000 applicants about hiring progress. Rather than playing the Blame Game for being ghosted by some company you applied to, just shrug and move on. Why spend any of your precious time stewing over it? Attribute it to the crazy job market and move on.

What's more, you need to stop blaming companies and their beleaguered, understaffed HR departments for tossing aside your resume if it is obvious you don't fulfill even the basic requirements for the job. There are job searchers out there that apply for hundreds of jobs per week. These are not jobs they are particularly qualified for or even particularly want. Some companies do have software that does some basic screening, but many do not. They actually read your resume and a very heavy slog it is, too. This is especially true of WFH postings.

To add insult to injury, I've read many posts where you, as a rejected candidate, want feedback as to why, exactly, you were rejected. As a hiring manager with potentially hundreds or even thousands of resumes to go through, that is a total nightmare. I would rather poke out my own eyes with a Sharpie. So would HR. A considerate company will send you a quick communication letting you know they've moved on with other candidates to allow you to move forward with exploring other opportunities, but as far as giving in-depth advice as to what is wrong with your resume or their impressions of your interview, they don't owe you any of that. They took up your time, but you took up theirs as well. Move on. It doesn't really matter why they rejected you. If you're a bad interviewee or your resume is lacking in some way, you need advice from an uninvolved party. Do some mock interviews. Ask some people familiar with your field to take a look at your resume. Remember that WFH opportunities mean a LOT of competition. You need to bring your "A" game in order to be successful.

To wrap this up, WFH is notoriously bad for growing your soft skills and providing the career growth you may want and feel you deserve. It is difficult to mentor or be mentored in a WFH scenario. When you have a question, what do you do? You probably zip off a Slack message or something similar. In return, you'll likely get an answer to your specific question. Nothing much more. That's a lost opportunity to actually have a conversation that morphs into something interesting and valuable to you.

Zoom meetings don't take the place of human interaction. They are normally meetings with an agenda and a group of people attending. That's not a good forum for any kind of honest discussion with either team members or your boss. Hopefully, you're having 1-on-1 discussions with your manager weekly. Those are normally Zoom meetings that can help the cause since there are only two of you. Overall, however, one hour of face time with your manager is not enough time for true mentoring and some of your best guidance and advice might be from someone who is not your direct manager.

When working from home, you can't absorb what is going on around you. You can't observe what kind of communications work and which don't. You can't suck up how people and areas interact with each other via office osmosis. You will remain oblivious to relationships (good and bad) and how those impact your work and that of your peers. Much of what we learn is through observation. Slack can't take the place of observation – too many office players are left out of those limited, stilted, written conversations.

Many WFH employees talk about feeling isolated and that's very true. You may be on a team, but you aren't really part of a team when each of you is in your own bubble at home and possibly in different parts of the world.

Your peers and your management don't get to know you very well. You don't get to know them very well. There's a limit to how well you can make friends and meaningful, potentially life-long allies and contacts through Slack or during a Zoom meeting. Business success is all about relationships.

Consider the problem of lay-offs. Contrary to popular belief, no company, anywhere, enjoys either firing or laying people off. Hiring people is a time-consuming, expensive proposition and letting people go leaves a hole in the organization and work that has to be redistributed or simply not done. It hurts people. Unless your boss is a seriously twisted mahambajamba, they will hate it. Besides the heinous task of letting individuals go, they are going to have to explain any departures in a way that helps them retain critical personnel and look strong to their public and customers. That means they may have to lie like rugs. Some will find that disturbingly easy and some will choke on it.

A WFH company normally has an easier time with this than other companies. It's much easier to let people go when you don't know about their families, their pets, their hobbies, or what's going on in their lives. It's easier to let someone go who doesn't make you laugh or who isn't friends with half the staff.

WFH can make you more of a playing piece than a person. The people you work with and for need to know you as a person. They need to like and respect you. They need to be able to rely on and trust you. You need people to care about and support you and you need to care about and support them. Everyone needs to be invested in the success of the company. Why? Well, for one thing it makes your working life a whole lot more pleasant. From a practical perspective, all of this helps you survive during hard times. What's more, it is going to help you for your entire working career.

People will forget projects they worked on 10 years ago unless there was something unique about them, but they'll never forget either great or hideous people they worked with. Your contacts can help you (or hurt you) later in your career and building meaningful relationships helps you, as a person and a valuable employee, grow. It's simply easier to forge connections in person.

I'm not saying your dog might not be better company. Hey, I'm not going to get judgy. I've been there, done that, and there were some situations where interacting with a toxic fungus would have been more enjoyable than working with a given person or team. Unfortunately, however, since it's unlikely your dog can get you a raise or promotion, learning to handle people and situations is a necessity.

Building a career isn't for sissies. In order to build a career, you have to work, you have to learn, and you have to either take or create opportunities for yourself. That means working with and learning from others. It is infinitely easier to do that in-person. You may need to sacrifice working from home if you want a meaningful, high-paying career.

That said, you might not care at all about a career. You might just want some work/life balance, enough money to be comfortable, and a job that allows you to focus on what you, personally, find important. Frankly, you are going to be a much happier person than someone who is ambitious. Ambitious people will have to work harder and sacrifice more. Regardless of where you fall in that spectrum, if you want to work, you'll need to follow the rules of whatever company you sign up with. At the time of this writing, the trend is away from WFH. Reality is a bitch.

Still, persevere in whatever direction makes sense for you. Realize, however, that working from home is moving in the direction it used to hold – an option for a few companies, but not the majority. Although companies are famous for Rhinestone Rhetoric (listen to what I say, don't look at what I do), they are not going to jump on board with this particular Rhinestone Policy until it can be proven that it offers some kind of significant financial benefit that is superior to the benefits of working on-site.

CHAPTER FOUR
ROCKING DIVERSITY...OR NOT

This is a hot topic right now and when considering the whole question of Rhinestone Quality, it deserves some discussion. Diversity is the epitome of a "rhinestone" concept – everyone says they want it or have it, but very few actually do. All that talk is smoke and mirrors. Most of the marketing flung at you, the customer, is carefully crafted to give the impression of diversity. Some of the pictures you see aren't even images of actual (or current) employees. They've been purchased.

Seriously? This couldn't be more obvious. How often do you think real companies happen to get together (they might all be remote), with one black, one white, one Hispanic, and one Asian (at least one of which will be a woman) all laughing and collaborating together? How many are tattooed, like, everywhere? Overweight? Older than 50, maybe holding a cane? Likely none. Those images are carefully crafted to tell a story. It's a pity that story is often a work of fiction.

This same kind of deception is often practiced at any type of newsworthy marketing opportunity. The company will put on a display of its diverse team members, whether or not those people have any business attending the event. It's all about the look, not the reality. The only positive thing I can say about these deceptive and disgusting practices is that if you're one of the minorities, willingly acting as the "face of the company" and allowing yourself to be trotted out and put on display can be beneficial for your career.

I'm going to talk specifically about IT, as that's my field, but inequities exist in every field.

On a scale of 1 to 10, IT scores about a minus 7 in terms of DEI. I was asked a few years ago to present something to motivate and encourage girls to look into and focus on my field as part of a STEM education program. I thought about it and finally refused. I cannot and could not, in good conscience, recommend the IT field – any of it – to an innocent girl child. It is a tough field period, but to climb that mountain as a woman means to you have to be beyond tough. The only reason I've been able to achieve anything in this field is that I am stubborn as hell and QA is simply where I belong and what I love. That said, I have mentored many women in the field. I simply would not recommend it to an innocent.

Let's talk about DEI in general first. DEI stands for diversity, equality, and inclusion.

Huh.

Everyone has prejudices. Everyone. If you think you have no prejudices, you need to sit back and self-assess. Prejudices exist everywhere. Prejudices don't have to be the obvious areas of color, sex, or religion. They might be educational (like having a STEM degree or other educational certifications from very specific sources). They might be personality-based, like having a decided partiality for people that display certain personality traits or abilities you personally consider particularly good. There have been tons of studies done that prove people have a tendency to hire people that are like them. So if you're an obnoxious asshole, you'll likely hire more obnoxious assholes. It doesn't matter what your color, age, sex, origin, sexual orientation, religion, disability, or anything else might be. You have prejudices and you should be aware of what those are so you don't make stupid mistakes.

I can't speak for other fields, but in my own, based on my own 40 years of experience, here's the breakdown of people in the IT field, in order of the most common to least common with a DEI focus – something our DEI programs supposedly helped and did not:

1. White men, straight, <50 years old (most in 30s)
2. Indian men, straight, <50 years old (most in 30s)
3. Black men, straight, <50 years old (most in 30s)
4. Asian men, straight, <50 years old (most in 30s)
5. White women, straight, <40 years old (most in 30s)
6. Indian women, straight, <40 years old (most in 30s)
7. Black women, straight, <40 years old (most in 30s)
8. LGBQT+ men, <50 years old (most in 30s)
9. Asian women, straight, <40 years old (most in 30s)
10. Workers over the age of 55
11. Hispanics
12. LGBQT+ women (any)
13. Disabled people (any)

Religion has never been much of a differentiator in any company I've worked with or for. But color, sex, orientation, and age has mattered a lot.

DEI has been around since 1965. THIS is how far we've come in 60 years.

Pathetic.

But what does this tell us? Not as much as you'd think. Clearly, DEI in and of itself has not been effective. In some cases, it has merely ensured that some companies have hired token personnel and used them to put a face on their alleged commitment to those principles. Nothing like using DEI as a marketing strategy, right? Once again, this is a Rhinestone Philosophy. "Look at us. Aren't we diverse??". Um, not really.

At its very worst, DEI can force a manager to hire for all the wrong reasons. Take a look at the differentiators between people. None of it has a thing to do with your skill set, your dedication, your experience, or your "worthiness". In short, nothing on the DEI list actually matters. Nothing. In a perfect world DEI would not exist. We wouldn't need it because not a thing on the list actually makes a difference in terms of doing a job.

But here's the painful part. We don't live in a perfect world. DEI was an attempt to address that which was and is the result of an unfair, prejudiced world. While I do not believe DEI has been successful, what would our world look like without it? Personally, I think it would be horribly similar to what we have now, but perhaps the programs were more successful in other, non-IT fields. Maybe things would be even worse. We are likely to find out. DEI programs are being phased out in the United States.

One of the reasons involves reverse discrimination. This includes hiring less qualified people for something other than their qualifications for the job. On the hiring front, hiring less qualified people for a job using something other than their qualifications happens all the time. As I said before, there is a decided tendency to hire people you relate to – who are like you. That is just a human condition and you have to be self-aware to combat it.

Those things can and often are personality-based. The majority of us are not self-aware. If you want to start to understand your own prejudices and are in a hiring manager capacity, take a look at who you've hired and why you hired them.

So the hiring process can show you your own failings, and the easiest person to change is yourself. You, personally, can change the world if you're willing to self-assess and work at it. My teams have always been diverse from a DEI perspective. The things DEI is supposed to promote have never mattered to me one way or another. I still have prejudices, though. I have a weakness for people who have a sense of humor, ambition, and a ton of practical experience. That doesn't sound too bad on the surface, but it is just as bad as any other prejudice. I can pass by someone who is traumatized by the interview process, has a ton of potential but not much experience, and so on. I actively work at it. You can work at it too. Become a better person.

And here's the thing – companies do not really demand anything better, so what we've got is a bunch of managers hiring little clones. Team members who are all the same are not necessarily going to be successful. They will all have the same strengths and weaknesses. What you need are team members that are complimentary. One person is good at something their team mate is not. A great team recognizes and appreciates what each of them brings to the table. One might be disciplined, organized, and experienced. One might be brilliant and wildly intuitive. One might be highly technical. One might be customer-focused. All of them should be competent. When you put those different people together, you get magic.

Nothing about diversity is pleasant or easy to write about. I've been a woman in a primarily male field for 40 years. Every heinous, crappy, disgusting thing that can be done to a woman in my field pretty much has been done to me. I'm not going to go into detail on those specifics – this whole chapter is depressing enough without it. I'd end up in a fetal ball sobbing into a hanky. The reason I can talk about these things, however, is because I've experienced these things. I often think the reason I'm qualified to mentor or teach is because I, as a newbie, pretty much did everything wrong, was terminally naive, and had to learn everything the hard way.

To make things even more difficult, we live in a visually-oriented society. We look at our world and each other and we make assumptions – some of which are really, really sad and very much off the mark. As thinking, reasoning human beings, you would think we had evolved further than that. Not. It flavors our interactions and in some cases it can hold us back. At the very least it stunts our own growth, personally, professionally, and ethically.

I'd like to talk about ageism specifically for a while. To be honest, I didn't even realize ageism was a problem. Until (guess what!), I reached a "certain age". And I find it very comforting that the people who are treating older workers badly right now are going to reach a "certain age" too.

Unfortunately, there is no handbook available on Becoming a Crone. Or Geezer 101. The changes are not especially fun, even if you take employment opportunities out of the picture. I'm not even going to get into it, as why spoil the surprise if you're not there yet? Overall, I can tell you it pretty much sucks Big Time. But what it doesn't affect is your experience, talent, or your mind unless you contract some sort of disease. In IT, most of our workers aren't exactly body builders anyway. We sit on our lard asses and work in front of a PC all day long. Why does age matter?

Well, consider how companies start, who owns them, who runs them, and the makeup of their core teams. Then consider our visually-oriented society and our prejudices. When you have teams of 30-something males, even if you, as a hiring manager, are not an ageist, you might hesitate to hire or promote someone that looks like one of your parents. Or (heaven forbid!) one of your grandparents. You might worry that person wouldn't be a "good fit". This will be especially true if your grandparents are sweet, old people in a home. We tend to categorize each other based on looks. The problem is that gramps there might be the Best of the Best in terms of your field.

To top it off, your young teams would benefit from working with highly experienced people who appreciate their craft, you, your company, and who do not have the same viewpoint as a relative newbie. In other words, you add to the understanding and capabilities of your existing staff. This is true of hiring ANYONE with that "might not be a good fit". What kind of teams are you building if someone who is a different color, sex, etc. won't fit on the team? How can your teams understand and appreciate your customers, unless your target base is exactly the same as they are? It inhibits your growth.

I have, in 40 years, worked for exactly one (that right, ONE) company that was diverse. Truly diverse. I can't even explain to you the richness of the culture. From a business and profitability perspective, they are #1 in their niche. The employees are devoted to the company, to their management, and to each other. Are they perfect? No, of course not. They have other problems. But diversity is something they got right and they profit from it every day. They are headed up by an inspirational, brilliant man, his leadership team is the same, and his company and the people in it are special in ways I can't even explain. That's what true diversity does for you.

You can build this too. Self-assess and work on your own prejudices. Look at your peers and employees and help them see and work on their prejudices. As a company, identify the areas that have diverse teams, particularly if they are high-achievers. See what they are doing right and take a good look at the hiring managers of those areas. Promote and reward them and use them as role models and examples to show others on your teams what you want to achieve. If you don't want to be a rhinestone company, you need to reward and celebrate those things you want to promote and work at what you want your company to represent. You need to walk the walk. It might be harder, but it's also more lucrative and gives you the kind of "real" reputation that brings you better candidates, better employees, and acts as a differentiator in the field. You don't need DEI to achieve this. You need your best selves.

This is also true of employees. I see a ton of stuff written about companies and their DEI journeys (good and mostly bad). But I don't see any stories whatsoever about how prejudices hold you, as an employee, back. We've talked about being a visually-oriented society. We look at each other and make assumptions. You simply can't operate that way and be successful in your life. There are good people. There are people who are Evil in Human Form.

None of it has a thing to do with DEI. There are evil people of color. There are heinous LGBQT+ people. Anyone can be an asshole. Every white woman over 40 is not a Karen. Every dude with tattoos is not a gang member doing drugs. And on and on. We need to extend each other some grace. If you, like me, have been treated poorly somewhere or by someone, why would you want to extend your own experiences and prejudices to others and become what you yourself most despise?

For example, if you manage someone talented, you may want to target them for a raise or promotion – or something they, personally, would find valuable that would let them know you see and appreciate them. First, you have to know them well enough to understand what motivates them. That means both of you have to get to know each other and be comfortable talking about those kinds of things. This is more than just your manager's responsibility. It's YOUR responsibility. Communication is a two-way street. If you avoid talking to your managers or coworkers for (whatever) reason, don't be surprised if they don't "get" you. You haven't given them an opportunity to "get" you.

If you've been targeted for promotion, your boss is going to want to mentor you and help you to succeed. Bad bosses are not going to deal with problems if they can avoid it. Most people aren't especially confrontational, but problems don't usually just solve themselves and go away.

Say you have been passed over for promotion a few times in favor of people less qualified than you are. If you have the nerve (I cleaned that up for you) to ask, your boss may or may not tell you the truth. It's going to depend on your relationship. If you dislike them or avoid them under normal circumstances, they're sure not going to feel comfortable talking to you about hard things. Like you smell bad, you don't play well with others (or others of a certain type), or something else that might be hurtful. This is feedback you need to be successful.

Say your boss or one of your peers gives you feedback that stings. It's highly unlikely they enjoyed doing that and it means they care more about helping you succeed than their own comfort. If you make the assumption it's because of your religion, affiliations, or your pink hair, you're doing yourself (and them) a disservice. Especially if you respond with something during the discussion indicating you think they're just saying that because you're (whatever). The problem here is not on their side. It's on yours. You've gone into denial that is going to ultimately hold you back because YOU are prejudiced.

It is always easier to blame others than to blame yourself. If you've been hurt in the past due to prejudice, I get that. But if you're going to get ahead – in anything – you need to become the person you wish others could be. If you're just as prejudiced as they are, but in the opposite direction, you are not part of the solution. You're part of the problem. Only assholes denigrate an entire group of people due to the actions of a few. Don't be one of them.

Let people help you. Let people mentor you. Throw away anything that doesn't fit into what you personally want to be, but think about and learn from your experiences, good and bad. Chuck your own prejudices and build relationships, allies, and friends with all types of people. It will change your life and you, as an individual, can help change the world.

Another reason it's important to talk about diversity, equality, and inclusion is if you're a manager in a rhinestone company that doesn't really value any of it, you need to be aware that you are going to have to advocate at least twice as hard to get someone who is different the recognition, raises, and promotions they deserve. Be persistent and force yourself to push through whatever roadblocks are put in your way. I can guarantee the end result and the satisfaction of being able to do something good for someone who deserves it is one of the most rewarding parts of management.

Considering DEI has been around for 60 years and left us with a big bag of nothing, we need to sit back and think about what comes next. I don't have the answers. You likely don't have the answers either. Everyone in a hiring manager's role, however, can change their little part of the cosmos. Everyone. You don't need a federal law in order to do the right thing and make yourself, your teams, and your company a role model for others. That much is in your hands. Go forth and make it happen.

CHAPTER FIVE
ON CORPORATE QUALITY

In IT, in particular, "quality" usually refers to testing and ensuring products are relatively bullet-proof.
Whether buying a coffee cup or a piece of software, "quality" to a CUSTOMER, means it does what you expect it to do and it doesn't break. That is the lowest common denominator in terms of quality. Simple.

In reality, however, quality is relatively complex and involves everything that touches the customer and a few things that don't. If you really want to be known as a high-quality company, your quality program needs to address all of it, especially if you have competition. Let's take a look at some of these things.

Let's start with what your product does, who it serves, and who is in your market. The time and money you spend surveying and getting to know your target market is critical. Consider, for example, buying a car. What a soccer mom and a racecar driver want and need are very different. If you do not correctly identify your market and build your product(s) around that, it doesn't matter if it works and doesn't break from a software perspective. Your target market won't buy it and the resultant lack of sales has nothing to do with IT. You did not identify and correctly build for your market.

What if you produce games or some sort of entertainment? There are additional levels of complexity here. The look, the feel, and the overall creativity of the app all matter here. You need to engage the imagination and entertain the customer and that's much harder to quantify. There's a huge difference between producing a game application and producing banking applications. When trying to deposit a check on-line, I don't want it devoured by a world-destroying monster I must then track down and kill before my transaction goes through.

Marketing research is a specialty in its own right and Marketing, overall, plays a huge part in how your company is regarded in terms of quality. They need to play a pivotal role in software development.

SDLCs (Software Development Life Cycles) and your company's normal operations, particularly those who consider themselves agile normally, tend to leave go/no-go decisions to the team, allowing unfinished work to be deferred to a later sprint or determining whether to delay a release to Production. Or they expect QA (their testing team) to act as a gatekeeper. A gatekeeper is someone that decides whether something moves on or not.

Often, in reality, QA is ignored. Not only ignored, but often denigrated, pressured, and insulted to boot. No matter – it has never really been a good idea to put QA in gatekeeper roles anyway. Their job is to find and report. They inform the team in regards to what is and is not working as expected. Although they may have a good feel for what the customer wants and will advocate accordingly as best they can, they did not make the decisions as to what best meets your customer's needs and do not know overall what will or will not be acceptable to the customer. While significantly more customer-oriented than developers on the team, they are still technologists.

Technologists tend to make the worst go/no go decisions on the team. Quite simply, they do not understand, use, or view the system the way your customers do. They often base decisions on what they, as technical people, can either get done on time or what appears to be important from their delightfully skewed viewpoint. Do not expect your technical staff to understand your customers (unless your customers are other technologists!).

Their job is to build what someone who DOES understand the customer has requested. Some of your technical staff have likely never even logged in or used your apps end-to-end from a customer perspective. This is not, by the way, a fault in and of itself. Your technologists will be focused on very specific and difficult tasks. You've heard the phrase "Can't see the forest for the trees"? That applies to your technologists, which means decisions as to what is and is not good for your customer should be made by people that represent and understand your customer.

Enter the Product Owner. A product owner is someone who translates what Marketing has discovered and, often partnering with UX designers, translates customer requirements into something the IT team can build. They may demo your apps or systems to customers. They may know many of them by name. They "own" the product, they show off the product, they talk about the product via conventional and social media, and may even be the "face" of the company to the general public, or they partner with Marketing to get that done.

When QA finds an error, it is the product owner who should tell the team whether it is something that must be fixed or something that can be deferred to a later time. They, and they alone, should understand what is critical from a customer perspective. The rest of the team does, of course, have a role as well. If something cannot be fixed or completed on time, that needs to be negotiated through and understood by the Product Owner, who (by their very title) "owns" the product.

If some of the work to be done in order to support customer requirements is technical, the Product Owner has to be sufficiently tech-savvy to understand the need and criticality for the overall product. I've met some genuinely brilliant product owners. They make or break it for a company. If you do not have product owners, perhaps you have business analysts or some other person that fills that role. If you have no one that fills that role, I strongly recommend considering getting someone and making them the gatekeeper for the team. No one else should be responsible for understanding and making your customers happy. They have other critical tasks and do not know your customers.

While we're on Marketing, let's talk about other aspects of quality from a customer perspective. Everything that touches the customer informs them in terms of the quality of your organization. Everything. Your website. Your blogs. Anything you post on social or professional media. Emails. Texts. And (oh yes!) Customer Service.

Your website is a primary Marketing tool, and a such, time should be spent to make sure it reels your customers in. There are courses available on good web design from a customer perspective and those designing your website(s) should know what attracts customers so they can avoid turning them off. I've taken a number of those classes – if you don't have graphic designers or anyone on your team that specializes in the look and feel of your apps, at a minimum make sure the staff working on this are educated in this area. The biggest problems I've seen on websites are typical for something that is technologist-designed. They are not visually appealing, the flow from a customer perspective through the features is cumbersome, they are slow, and they simply have too much stuff on the pages. In other words, they are over-designed.

I don't want to have to read a novel to do something simple. When you make updates to your website, have someone who was not involved with the updates check through it. This is a basic QA task – someone other than the person who creates or updates something checks it before it goes live. Save yourself some embarrassment or a website that is unavailable to your customers for a period of time. There are companies out there, by the way, who offer services that constantly check your website to ensure it is alive and well, but it is a relatively easy task to code up something yourself, if you are so inclined.

The same basic QA task should be applied to anything that touches your customer. Don't publish a blog until someone else on the team reads through it. Have someone look at anything for social or conventional media before it gets published. Have blast emails or texts checked by someone other than the author. This is so simple and does not take much time or effort, yet it is constantly ignored, which can cause the company considerable embarrassment.

It is a weird part of human nature that we tend to miss our own errors – repeatedly. Consider if you write an important report for the CEO. You re-read it a hundred times, find some mistakes, correct them, and eventually you're convinced it's solid. Then you ask a colleague to take a look at it. They find 5 mistakes right away – some of them glaringly obvious. How did you miss them?

Welcome to being human. We just do not see our own errors even when we use tools to help ourselves out. That's why QA –

an independent 3rd party – looking at and verifying your stuff is important. QA applies to everything that touches your customer.

Moving on, a major part of your quality landscape, since everything that touches the customer informs them as to your quality as an organization, is Customer Service.

You can have the best product available on the market, it can do everything one could expect and more, it can never break, and one bad experience with Customer Service will lose you a customer for life. Furthermore, it will lose you any customers that one person can influence for life as well. If you have any competitors whatsoever, a disgruntled customer will leave you in a heartbeat.

 It is critically important that a customer who calls you for help does not feel frustrated, insulted, or angry. What does that mean? It means if you have a phone system that largely takes the place of your customer service, (press 1 for Department X…), the minute the customer hangs up in disgust or has to hang on the line for over half an hour, you've lost. In order to use Customer Service, you have to be able to reach Customer Service.

The importance of great customer service is hugely under-rated. A friendly, empathetic, knowledgeable human being that speaks your language and can solve your problem is like solid gold. They should be (and usually aren't) paid accordingly.   These people know your product. They know what makes your customers crazy. They are your first line of defense when you release a product or update that has or causes problems. Are your customer service support personnel kept up-to-date about what is in new releases? Are they kept up-to-date about known issues with the release? If not, you are hamstringing your entire company and potentially losing customers you could have kept.

Does your Marketing area make Customer Support their Very Best Friends? If not, they should. Does your IT department treat their problems with promptness and respect? If not, they should. Does your company draw from this pool of knowledgeable skill when openings occur elsewhere in the company that could benefit from that product and customer knowledge? If not, then you are wasting time and money sourcing people for jobs that don't hold a candle to what you've already got on hand.

You have to care about your customers. If you don't, you might be on top for now, but the nanosecond a viable competitor is available, you will lose them. I have to say it is disheartening and sad to be fed a line about being customer-centric when a company is money-centric and that's about it.

I can tell you right now I can tell how much you care by the quality of your customer service. I am English-speaking and I work in an English-speaking country. If I cannot understand the customer service rep and have to ask them to repeat themselves multiple times, you're a money-grubbing asshole more interested in saving a few bucks than solving customer problems.

If the customer service people can only solve a very specific short list of problems and I keep being referred to someone else, you haven't invested in customer service. If I can't find a number and a way to speak to a human being from your app or website and your on-line "chat" isn't useful, I'm going to be a frustrated, angry customer that seriously wants to be done with you. Is that what you want? Customer Support can be a differentiator. Think about it. Cheaping out on Customer Support is not a good idea.

Let's move on and talk about Sales for a bit. Sales fascinates me – probably because I am absolutely terrible at it. I can't sell something to someone when I sense they may not really want or need it and I can't sell a product I don't love myself. I'm somewhat of an introvert and have virtually no small talk. I'm just abysmally bad at sales. We can't be good at everything – which means I enjoy and respect those who excel at what I do not. I've heard tons of criticism about sales people. Selling your product – whether through the efforts of your Marketing or your Sales staff (or more likely, by those areas working together), is how your company pays the bills, expands, and thrives. You need to sell your product. A great, appropriately compensated sales team is critical to the company's success.

Quality plays a role here too. The best sales people I've ever known not only come across as strongly believing in their company and products and they are totally convinced their solutions are going to help you. That you genuinely need their products. Their sincerity shows in their voices. People who are successful in sales need this sincerity (whether it's real or not), they need to enjoy talking to people, they need to be persistent, and they need the hide of a rhinoceros. If that deal they were working on for months doesn't go through, they need to toss it off and move on with the same level of belief and enthusiasm to the next potential client. I would likely fling myself off the nearest bridge after several adult beverages.

Where quality plays a role here is there needs to be agreement in advance as to what your sales people are allowed to sell. The tail can't wag the dog – your company cannot have to constantly galvanize and exhaust itself fulfilling a promise made by a sales person to an important potential client that they had no right to make. Nor can they commit to dates for new stuff the company as a whole has not committed to or made public.

The reason is simple. If you promise something you cannot deliver, you have just lied to a customer. Maybe not deliberately, but they are unlikely to be understanding about it if they have invested money in your solution. You will have difficulties selling to them in the future if they know it's all smoke and mirrors.

Customers talk to each other. Through conferences, through conventional media, and through social media. To be successful, most companies need to be customer-centric. That doesn't mean there aren't successful companies out there that are unethical, thieving bastards. But it does mean their future is iffy, no matter how big they are at the moment. No one stays with someone that treats them poorly if there are equally good choices available in the market.

Another benefit from a good sales team is their ability to help inform Marketing about trends and customer requests in order to help set the direction of the projects the company chooses to support. If they lose sales due to a lack of some feature or product, their feedback can help the company course-correct.

Sales people, particularly for large companies, can make a ton of money and receive all kinds of sexy incentives for making or breaking sales quotas. It can cause resentment in other areas of the company that make or support the product they are selling. But it shouldn't. Selling your product is the lifeblood of your company. Keeping those customers happy ensures your company's continued health. That sale they just made might fund your promotion. It certainly helps the company pay your salary.

Much of what you do and how you represent yourself to the general public shows the world whether you are a rhinestone company or the real deal. If your marketing efforts talk about your quality and reliability – followed by making the news over a recall or failure, people might still buy your product or service if you don't have much competition, but they aren't going to believe a word you say and what's worse, when the news (which likes to makes things as sensational as possible) throws you under the bus, everyone is going to believe it. There's a tendency to believe Big Business is inherently evil. So if quality and reliability are part of your branding, you need to believe in it and invest in it.

Whatever you represent as to who you are and what you do to the general public needs to be true. So if it isn't true, either invest in making it true or find other good ways to describe yourself to your public that represent something you can really get behind and execute with pride.

The same is true internally.

When I navigate to your careers page, do you tell me what your values are? Do you operate in a way consistent with those values? For example, you may say you value integrity, innovation, and excellence. All of those are nice words. If, however, your staff work under conditions that squash innovation, you do not celebrate and reward excellence in a way your people understand and support, and do not show any integrity, your values are a lie and you are a Rhinestone Quality company.

If someone asked one of your employees about those values, they would either laugh bitterly or shake their head and say "Yeah. Not.". Your values and your mission need to be more than weasel words. They need to be something your executive team believes in, supports, invests in, and promotes.

Your people should be proud to work for you. They should feel some loyalty to you and difficulties forcing themselves to leave or consider greener pastures. They should brag about you. Do they? If not, why not? Those are things you need to work on.

Your customers should like you so much they don't want to go elsewhere. Do they show any loyalty to your brand? If not, why not? Those are things you need to work on.

If your answer is no to those questions, you need to take a hard look at your culture, your stated values, how you actually operate, and how any disconnects have impacted your bottom line or could jeopardize your future.

As an employee, you need to ask yourself if you are contributing to the company's success (which impacts your own) and supporting their values. If not, you're expendable.

Corporate quality is everything that touches the customer. What you do as executives, as managers, as employees, all potentially informs the rest of the world. Everyone needs to make their part of the puzzle count.

CHAPTER SIX
WE'RE OFF TO SEE THE WIZARD....

No book about Rhinestone Quality would be complete without talking about Thought Leaders. Influencers of the business world, Thought Leaders are regarded as the intellectual brainiacs of our field, perched securely on top of the mountain, dispensing wisdom to ignorant, lesser mortals such as ourselves.

I don't regard myself as a Thought Leader. I'm a practitioner. I'm going to be honest here (why not?) and admit that I've had a Whole Lotta Fun during my career poking at Thought Leaders. I can't help myself. I'm not much of a follower, and not likely to join any clubs or cults. I really appreciate new ideas, providing I can kick the tires, play around with them, and change them around so they are practical and useful for whatever makes sense in my current situation. I have a positively wretched sense of humor, have been in a field where cynicism is both a survival mechanism and an art form, and I strongly dislike people or movements that denigrate others in order to make their points. It stands to reason I take Thought Leaders with a grain of salt.

That said, Thought Leaders did not attain their lofty status without followers that strongly believed in what they had to say. Many of their ideas and beliefs are on-point or at least interesting. It's up to you to sift through all the rhetoric and find some diamonds you can actually use.

I thought I'd start out with a simple definition, but the definitions were actually more complex than I imagined.

What is a Thought Leader? The most common definition is someone with deep expertise in their field that shares their knowledge and ideas with others. Wow. That makes most of the people in the field at a senior level or above into Thought Leaders. It's similar to the phrase "Everyone is a Philosopher.". It's also true. If you're an expert at what you do, you share your knowledge, and help others, you too are a Thought Leader.

Unfortunately, that definition doesn't really cover what most of us think of when someone uses the phrase "Thought Leader". I dove a bit deeper. There is a definition from Forbes that tickled my funny bone, resonated with me as true, and I'd like to share with you now. "A thought-leader is a person or organization that benefits financially from being regarded so.".

Bwahahahaha.....

Let's think about that. What that statement says is a Thought Leader's motives aren't altruistic and they benefit in some way financially if you believe in them. Puts a different spin on things, doesn't it?

Consider the hype and panic over what is happening in the workplace right now. Articles from respected sources are popping up everywhere helping to fan those flames higher and feed into your paranoia. Fingers are being pointed at those you're most likely to blame anyway, as most people are happy to believe everything wrong with the workplace today is due to rich, entitled bastards and evil companies. What's worse, there seems to be some belief that it's a new phenomenon. It isn't.

Have you actually clicked on those articles? They want to sell you a subscription. In reality, there isn't much that is actually either new or useful in those articles.

Do you really think the phrase "No one is irreplaceable" is new? It isn't. It's over 100 years old. Do you think AI is the only technology ever designed that scared people into thinking their worlds were ending? It isn't. Those articles are out there to allow you to feed your belief systems and to sell subscriptions.

Whenever you're going to listen to someone as an expert, you have to take what they have to say in context. What do they have to gain by swaying your support their way? In the above example, it's selling subscriptions. In other examples, it would be to get you to buy their tools or services. You may find genuine value in any one of those things, but you can't be naïve about it and just believe everything you read or see. There are a lot of cheap rhinestones out there and not many diamonds.

Does what is being said have some practical application in your reality? Is it just a declaration of a problem, or does it offer some solution to the problem? Why is it a problem in the first place? To use an example in my particular field, if you just say there were 362 studies done (with no reference as to what studies or by whom) and manual testing sucks, the first thing I'm going to do is look at what you do for a living. If your company sells automation tools or AI services, you have a vested, financial interest in getting people to see things your way. That doesn't mean we shouldn't listen to what you have to say; it might be interesting and informative. Believing it as gospel, however, is another story.

Where is real-life proof of your claims? Your solutions should have been of benefit to some company other than your own. If you throw numbers at me, those numbers should have some meaning. Something like the number of tests created or run is meaningless. What matters is running the right tests, at the right time, in the right environment. It doesn't matter if the test was manual, automated, or generated by AI. Was the end product better as a result? Prove it. Was the process faster? Cheaper? Prove it. Convince us that rocking our companies' worlds and handing you a pile of money is going to benefit someone other than yourself.

For those in the quality field, are we quality professionals? Are we practitioners? If so, it means what sets us apart should be a natural skepticism and a burning need to understand things, which leads us to ask a lot of probing questions and test the crap out of anything that comes our way. That needs to include the bloviations of someone trying hard to set themselves up as a Thought Leader. It does not matter who said what if it makes no sense in your reality. Actually changing your reality is hard. You may find the juice isn't worth the squeeze.

So be critical.

Do you have problems with rants that are purely philosophical in nature? It might be interesting and it might not, but philosophy, in and of itself, is not going to solve a problem. It might define a problem – at least if it is accessible. Ideas need to be approachable and easily consumed. You can't move the field forward, back, or sideways if the majority of your audiences don't understand you.

Look, you may be a genius. In fact, like Wile E. Coyote, you might be a "Super Jenius". Your closest friends might be geniuses. Even your goldfish might be a genius. But guess what? Most people are normal. In fact, some of the people you might need on your side may even be somewhat challenged in the intelligence department. That is just reality. Do your ideas help them succeed? Do they understand them? If not, toss those ideas. Rethink them. If you really want to make a difference, benefit the majority rather than the gifted few.

Let's address what it means to be normal for a few minutes. Normal is good. As someone who has been in managerial roles for a long time, I'm going to tell you something shocking, controversial, and borderline unbelievable. Unless, of course, you have also been in a managerial role for a long time.

Geniuses are a pain in the ass.

It always makes me laugh when a company says they want to hire all 'A' players. Why? Because you can't accomplish much with all A players. They are picky about what they work on, how they work on it, take instruction poorly, often take criticism even more poorly, may have trouble working with their normal team mates, and question everything.   They suck up at least three or four times more attention than the rest of the staff. Do you really want an entire team of geniuses? If so, I freely tip my hat in your direction. You are a better person than I am.

I have been lucky enough to work with a number of geniuses. They are a welcome member of a team. But an entire team? Or worse, an entire company? Argh. Can you say "nightmare"?

'A' players are going to want to work on those things that interest and challenge them specifically. That means you are going to have to hire and partner them with disciplined, competent people to work on all those things that bore and frustrate "A" players. Otherwise, you're not going to get anything done. Not everything we've ever had to work on in our lives has been interesting, ground-breaking, or exciting. You'll need people who are accepting of eccentric, quirky, and occasionally arrogant people that sometimes aren't very good with other human beings. When you're always ten or twenty steps ahead of everyone else, it's really hard not to regard others as somewhat stupid, which makes it that much harder for geniuses to build relationships and have satisfying work environments.

So if our Thought Leaders are geniuses, does the same apply? Generally, yes. They may not be very understanding, kind, or capable of using common sense. They can be bombastic, condescending, and unrealistic, as they have genuine difficulties understanding the reality most people work with every day. What is obvious to them is not obvious to others. I'm sure it frustrates the hell out of them. Unfortunately, as I'm the irreverent type, I can't bring myself to care all that much. Genius needs to be accompanied with a healthy dose of common sense.

That said, watching Thought Leaders of different camps and beliefs attempt to annihilate each other is extremely entertaining, especially if you aren't in the middle of it. There are times I've wanted to sell popcorn, put them in a mud pit, give them nerf swords, and watch them beat on each other. Unfortunately, their sparring is never physical, although I think I've seen frothing and foam flying. Watching Thought Leaders trying to verbally rip each other new orifices is always exciting and fun. For you, not for them.

That said, if you're going to poke at a Thought Leader, be prepared. If you hit a nerve, question their ideas, or present anything that could be viewed as a threat to those ideas, gird your loins. Further, if they have a cult-like following, things can get a little scary. I've had death threats. No, I'm not kidding. Never from a Thought Leader, though. Their followers, however, can get a bit...passionate.

Are all Thought Leaders this way? Hell, no. All of us can be Thought Leaders, remember? I don't identify that way myself – I've never made my living or supported any kind of empire from selling my ideas to the field. My whole career has been doing very specific work for very specific companies. Again, I'm a practitioner. As jaded as I am, however, I am not "against" Thought Leaders. We all need to learn from each other and from our field. We all have our favorite Thought Leaders, and I have my favorites too. People I like and respect. In my case, it's always going to be someone sane, who has something to say that is relevant to me and where/how I work, and frankly, someone with a sense of humor about their world. There are Thought Leaders out there that make me laugh all the time. They are all kinds of twisty inside.

I took a class from someone who was a Big Name at the time and can honestly say it changed my entire career. Not because I slavishly applied every golden word to my situation, but because it got me thinking.... I took what I learned, hacked it up into weird pieces, massaged it quite a bit, and came up with something that not only changed the way I taught and set things up, it shaved weeks of time off our schedules, and to date has benefitted at least five companies I've worked with since. Speeding things up with no loss in quality? Oh yeah.

That, my friends, is why it's important to get out there, listen, and pay attention to what is going on in your field. It's highly unlikely someone is going to recognize problems unique to the company you work with and magically have a solution to all of those problems. When you hear claims like that, there is no harm whatsoever in sitting back and listening. Then go back to work and play around with those ideas if they seem promising.

Part of what is important to you and your career (and your company!) is staying relevant in a world that is constantly changing. I'm convinced I've been in the IT field so long because I'm easily bored and the IT field, in particular, is all about change. If you are resistant to change, it's a poor career choice.

We've probably all experienced change in a variety of ways, but the best kind of change is the type you orchestrate and control yourself. Let's use AI as an example, since there's such violent and opposing viewpoints about it at the moment. Let's say you are absolutely opposed to the idea of AI doing the work of human beings. That's fear talking, by the way. We fear what we do not understand. Something shuts down in your head and you refuse to entertain the idea of using AI (for anything, but particularly in the workplace) so you don't go out and learn about it. By the way, I've seen these types of scenarios play out for many other "new" technologies and ideas. Automated testing was viewed that way as well. So was Agile methodology.

What is eventually going to happen to you is that what you most fear and know nothing about is going to be forced upon you. It will be forced upon you by people who do not understand the work you do or how you do it. It would have been infinitely better if you had expressed curiosity and excitement, gone out there and learned everything you could, played around with some tools, and gone to your management with your own suggestions and recommendations, backed up with your experiences.

In other words, you can own the change or the change can own you.

Someone who keeps on top of what is happening and what is possible, advising management accordingly, is viewed as innovative and valuable. Someone who just says no is viewed as an impediment. Do you want to be a hero or an obstacle? Do you want to fit new technology and methods into what you've already got, enhancing your company's products and efficiency? Or do you want some bogus piece of crap imposed on you that will take forever to get implemented, if it ever can be implemented?

That means when I go to (or send) someone to a conference, I consider it a rousing success if there are a few germs of ideas in there that might be useful. If you don't go looking for them, they'll pass you by. Go learn about things that scare you or that you don't understand.

One of the great things about learning is it genuinely doesn't matter what your position is in the company. Innovators are usually highly appreciated. What's more, if you experiment and learn, later sharing your findings not only within your company, but outside in the field as well – guess what? You have become a Thought Leader!

CHAPTER SEVEN
YOU'RE NOT AGILE

Do you work with or for a company that proudly proclaims it is an Agile Shop? Are you really agile? Have you ever actually read the Agile Manifesto or its underlying principles? If not, how do you know? Let's take a look at those things.

The original Agile Manifesto was relatively simple:

Individuals and interactions over processes and tools
Working software over comprehensive documentation
Customer collaboration over contract negotiation
Responding to change over following a plan
That is, while there is value in the items on the right, we value the items on the left more.

That's it. It first came out in 2001 and was a big deal in the IT world at the time. Everyone was lining up to sign the manifesto and "big names" in the field showed off their signing and support.

The entire effort was based on a small, tight team that pulled off a major effort by themselves and who later documented what they had learned and attributed their success to the adoption of some repeatable processes. Later, when they had been inundated with questions and requests for more information, the manifesto was expanded to include more detail regarding the principles behind their manifesto.
I'm including all of this here because it's a rhinestone philosophy when you say you do something you clearly don't do. These are the principles they follow today:

1. Our highest priority is to satisfy the customer through early and continuous delivery of valuable software.
2. Welcome changing requirements, even late in development. Agile processes harness change for the customer's competitive advantage.
3. Deliver working software frequently, from a couple of weeks to a couple of months, with a preference to the shorter timescale.
4. Business people and developers must work together daily throughout the project.
5. Build projects around motivated individuals. Give them the environment and support they need, and trust them to get the job done.
6. The most efficient and effective method of conveying information to and within a development team is face-to-face conversation.
7. Working software is the primary measure of progress.
8. Agile processes promote sustainable development. The sponsors, developers, and users should be able to maintain a constant pace indefinitely.
9. Continuous attention to technical excellence and good design enhances agility.
10. Simplicity – the art of maximizing the amount work not done – is essential.
11. The best architectures, requirements, and designs emerge from self- organizing teams.
12. At regular intervals, the team reflects on how to become more effective, then tunes and adjusts its behavior accordingly.

This methodology is not new – it's 24 years old. On the surface, all of the above sounds great. In practice, not so much.

I never signed the Agile Manifesto. I'm not much of a manifesto-type person. I've worked in some highly regulated, heavily audited organizations. I don't agree with a number of these principles. To be completely honest, Agile doesn't mean any more (or less) to me than Waterfall, RUP, or any other software development methodology.

In other words, I don't care.

If I wrote my own manifesto, is would be:

"A great team will survive, surpass, and succeed regardless of methodology and obstacles.".

That's it. Feel free to sign up. But back to Agile, and why you aren't. Maybe you wish you were agile. You kinda are – after all, you hold retrospectives, don't you? However, when you say you are something you're not, you're lying to yourself, your employees, and worst of all, you're doing a disservice to your own company and to the field in general. You're giving strength to the lie that a methodology that doesn't work is used by everyone.

From a corporate perspective, it isn't especially helping your bottom line, which is dependent on producing awesome things people want to buy as quickly and efficiently as possible, to say nothing of not wasting time and money on processes that are not of any fiscal benefit to anyone.

Everyone says they're agile, so you aren't alone. Everyone wants to be agile. It's the word itself. Who wouldn't want to be quick and well-coordinated?

But chances are good you don't really follow agile principles. Let's look at why.

If you are committed to AI or automation in terms of the bulk of your efforts in order to be faster and cut down on the number of people you hire or retain, you value processes and tools, which you likely equate to time and money and you do not value individuals and interactions above those things.

This is a natural tendency for technologists. Engineers grow up to be managers, directors, VPs, CTOs, etc. Talking about tools and processes push all their "good" buttons. It's what they've done, what they do and understand, and what makes them comfortable. They'll bring the rest of the executive team along for the ride when they mention time and money, whether that is actually true or not and regardless of the end result on their customer base. If you operate this way, and many do, it means you are already at odds with 25% of the founding principles behind agile methodology.

What about "working software over comprehensive documentation"? It depends. What do you mean by "working"? If it is something actually made available to and useful to a customer, many agile efforts fail to do that. If it means a piece of software that appears to work by itself in a virtual vacuum and which will be unavailable to the customer until the feature or application is more complete, that makes more sense. And what is "comprehensive"?

I'm not a fan of red tape and I'm sure you're not either. If comprehensive means "anything more than absolutely necessary", I can get behind this. But even "necessary" is going to differ tremendously between companies depending on the nature of their business. A highly regulated and audited field requires more documentation than a start-up in other fields. There are also companies that have political climates that are so vile they require a lot of red tape and sign-offs just to force people to work together and get basic work done. So overall, the concept of working software over comprehensive documentation is vague enough to be worthless.

To put a more positive spin on it, however, you can define this yourselves. It's your company and your environment. What is "working software" to you? What is enough documentation? What is overkill? And why even waste time talking about it? Because if you don't, every person you hire is going to have their own interpretation or if in a management position will try to impose their own understanding on you – wasting your time (and of course, your money).

Being efficient as a company means being self-aware and cutting the crap, making sure everyone understands and is on board with how you do things and want to operate and what you value. If you have upper management, laying down strategic guidelines is part of their job. So get strategic. Otherwise you'll be trying to manage a free-for-all. That can be fun, but it's also exhausting and not sustainable long-term.

I've always had problems with "customer collaboration over contract negotiation.". I wish this said "customer collaboration is as important as contract negotiation.". Your contracts establish too many things critical to your success as a company to take a backseat to anything. Poorly negotiated contracts can bankrupt you or get you sued. Bad contracts inevitably result in cost overruns and a loss of revenue. You can still be committed to customer collaboration without sacrificing contract negotiation. If you believe this as well, the concept represents 25% of the core values behind agile methodology.

Are you still with me?

Let's talk about "responding to change over following a plan". Wow. Many "plans" in terms of agile methodologies are a few weeks long and consist of pulling a number of tickets in front of the team to be "worked". There is no real "plan".

Significant, constant change creates chaos and ultimately translates into lateness on other things – some important and some less so. Constant change is the bane of many agile teams' existence. Their shops are chaotic, messy, and nothing ever seems to get done, de-motivating and frustrating many of the staff. Change needs to be intelligently managed so it doesn't negatively impact your progress or your teams. Every change should be evaluated and if it isn't critical, it needs to go to a later sprint (or phase) in order to protect the team and their progress.

Next to consider is "Our highest priority is to satisfy the customer through early and continuous delivery of valuable software.". I laugh every time I read that – it's such a total load of steaming male bovine excrement.

Does your customer want continuous delivery of software? Is it actually valuable to them, or valuable to you? How do you measure whether you satisfied your customer? This should read "Our highest priority is to increase our bottom line through early and continuous delivery of software that makes us more money.".

First, people as a whole do not respond well to change. It is unlikely anyone wants software that is constantly being changed, unless the change is a fix. In some cases, constant change can actually drive people away. You need to know your customer base and what is and is not acceptable to them.

Many companies deal with "early and continuous" by actually hiding new features and functions behind something like Split.io until a nice, marketed upgrade is scheduled with a bit of fanfare. Some make the upgrades available gradually, a few clients at a time. Regardless, the real driving force here is not the customer, except as a limitation. The driving force is an ever-increasing, ever-improving product base that delights your customers and thus increases your market share and value as a company.

"Welcome changing requirements, even late in development. Agile processes harness change for the customer's competitive advantage.". No one I know, and I know a lot of professional people, welcomes changing requirements, especially late in development. And agile processes do not harness change for the customer's competitive advantage. There may be no competitive advantage whatsoever to the customer. It's more likely the change is advantageous to the company's competitive advantage. And if it isn't, why are you doing it? Unless it is something being demanded by some of your largest and most influential customers, it is likely something that can be deferred to a later sprint, when it isn't disrupting your progress and your teams.

Generally speaking, it is a mistake to allow the tail to wag the dog. What a short agile process does allow you to do is to incorporate change more quickly – you can choose to place changes in later sprints without blowing what everyone is working on out of the water.

I once worked for a company that was virtually hamstrung by dropping everything to work requests from their largest customers. They ended up being used as personal IT departments by these customers and were making no progress against corporate goals that involved updates and new features for all of their customers. Eventually, operating this way could not be sustained – they had to stop. They lost some of those customers. If they had set boundaries earlier, it would have been less of a problem. This is especially prevalent with small companies and start-ups, who are more dependent on large, vocal accounts.

I believe a good team can incorporate change, but will avoid disrupting their work whenever possible. Ultimately, what you want is up to you. Is a chaotic madhouse scrambling to get things done at the last minute every two weeks attractive to you? It's certainly exciting. On the other hand, it's certainly the opposite of a well-oiled machine and there's a lot of burn-out. Human beings cannot maintain that level of excitement and terror long-term and after a year of that craziness, with the accompanying overtime and angst, your staff will cease to be energized and will no longer pull together to get things done. When everything is an emergency, eventually nothing is an emergency.

The next principle is "Deliver working software frequently, from a couple of weeks to a couple of months, with a preference to the shorter timescale.". I can only assume "working software" means a piece of code that appears to work by itself and that will be hidden from and not usable by a customer until the entire feature or function is done.

There is a huge difference between providing a fix and providing a new feature/function. Fixes can and should be made available as they are ready. New features, functions, or applications are marketing opportunities that you will likely want to introduce to your existing and potential customer base in a way most beneficial to your company and easiest for your customer base to accept and even look forward to, thus making you maximum money on your investment.

New features and functions might be extremely simple – something within the capabilities of a single team in a single sprint (or block of time). But a new application or some features will be bigger than a breadbox and will require multiple sprints to complete.

I've worked on efforts managed through agile methodology that took two years to complete. Progress was still measured in two- week increments, but the final product was not complete and offered to customers for two years. Take a look at that. This is one of the reasons methodology doesn't matter much to me. Two years sounds waterfall-y, doesn't it? But it isn't.

One thing agile is really, really good at is sweeping their failures to get work done in a timely manner under the rug. If something isn't done on time, it's moved into the next sprint/block of time. Essentially, development is a creative process. How long does it take to catch a fish? Paint a picture?

Things come up that were unplanned, things turn out to be more complicated than anticipated, in essence, shit happens. Agile methodologies are better at hiding how long it really takes to get something meaningful to the customer. I'm not sure whether that's a benefit or a fault. You decide.

Next on the list is "Business people and developers must work together daily throughout the project.". If "business people" is equivalent to the product owner, this can happen. If it means anyone else, it won't. Business people might be involved during the ideation and specification phases, but that will likely be it until there's a product out in the market. Often customers aren't involved at all. Look at your own company and situation. Do you have business people, developers, and maybe customers working together every day? If so, do you find that type of collaboration speeding up your process? Probably not.

I strongly believe in and support a few principles of agile methodology and the next is one of them. "Build projects around motivated individuals. Give them the environment and support they need, and trust them to get the job done.". There is nothing a motivated, talented, tight team cannot accomplish. Nothing. But this is genuinely one of the hardest of all the principles to get right.

First, do you have any teams you can definitively say are motivated? Or have you sucked the life out of all your talent? Or not hired the right talent? Have you built out the kind of environments they need to do their best work? Do you support them? Do you even understand what makes those team members tick and do their best work?

No matter what articles and studies you read (often to validate what you want to believe), many people are motivated by money or reward. That said, however, everyone is different. I have found that most people will knock themselves out for someone who listens and responds to them, believes in them, advocates and brags about them (especially in public), protects them, and just in general demonstrates that they care about them as human beings and respects them as professionals.

One of the key points in the agile principle is "trust them to get the job done". Trust is an unbelievably hard thing to come by in most companies. When you read the multitudinous complaints out there about managers, much of it stems from lack of trust and micromanagement. Hire the right people, give them what they need, and then get the hell out of the way and let them get the work done.

A really good team doesn't need you in order to do their thing. The team needs you to do things they can't do – and that shouldn't be their day-to-day work. If they are programmers, you shouldn't need to look at their code to make sure they're programming things correctly. If they're QA people, they don't need you to check their test automation or test cases. You're going to know if things are going awry by results, at which time you can step in and course-correct if necessary. Pair senior people or leads with those less experienced and let them gain experience with training and leading teams, as well as counting on them to lead through example. Encourage your people to step up. Train your people to step up. Believe in their ability to step up.

If you are not an "individual contributor" (actually, I hate that term, since everyone should be contributing something), then you should be working on strategic initiatives. Leave the tactical work to your teams. If you don't understand the difference, you have a problem. I can't solve that here. It needs its own book.

To move on, the next principle is "The most efficient and effective method of conveying information to and within a development team is face-to-face conversation.". Now be honest. Do you have co-located teams? Or do you have remote teams or some members that are remote? Then by definition, you cannot be agile. In spite of the fact that many of you will hate me for saying this, I agree with face-to-face conversation as the most effective means of conveying information. I also believe teams are more creative and innovative when they're collaborating with each other in person. But if you don't support that (or can't support that), you aren't really following agile principles.

Agile methodology supports working software as the primary measure of progress. I can buy that, since every software development methodology supports the same thing. The only caveat with agile methodology is what you define as "working".

Let's talk about "Agile processes promote sustainable development. The sponsors, developers, and users should be able to maintain a constant pace indefinitely.". Nothing I've ever experienced in regards to companies that say they are agile shops supports this.

Most companies choose Scrum (a type of agile methodology) to produce software and most choose two-week "sprints" (blocks of time in which to produce results). The pace that has to be maintained tends to de-motivate and burn out staff, unless the only work being done are fixes or tech debt.

Developers tend to be optimistic about what they can accomplish in two weeks. That means they have to push, really push, to meet their commitments in a two-week timeframe, often with overtime and a lot of urgency and anxiety around the deadline. They go through that every two weeks with Scrum. It is not sustainable.

Most human beings and teams do better work with some sort of deadline. It's a goal that keeps them moving forward at a good pace. But two weeks is usually an arbitrary and meaningless number; this is usually just pulling a date out of your ass for no reason whatsoever. Why would you expect good results from this?

We're almost through. Next is "Continuous attention to technical excellence and good design enhances agility.". That is an absolute lie. Technical excellence and good design take more time. It also takes more training. I strongly believe this is part of agile methodology principles because they were written by and largely for technologists and all of them want to incorporate this into what they produce.

Excellence, in anything, requires more time, money, and commitment than mediocrity. Do you have a lot of tech debt? Things you should have done but didn't have time to do? Then you don't adhere to this principle. Here's the painful part. Maybe you don't need to in order to be successful. At the very least, you can define what constitutes "technical excellence" and "good design" within the context of your own company and associated products.

The next principle is " Simplicity – the art of maximizing the amount work not done – is essential.". OK, I guess. Extremely difficult for technologists, though. Overdesign was, is, and will continue to be an issue for technologists. Fortunately, applications are usually designed by non-technologists. Any products crafted for use by non-technologists should be designed by non-technologists.

Keeping things simple is not really a new or strictly agile concept. If you take a look at what we actually have to work with out in the wired world, it's obvious that "simplicity" isn't necessarily a priority. In addition, consider how companies really work. You may design a product or a suite of products. Your customers love your products. They are the highest-rated products of their type in the world. And yet. And yet. You keep poking at them. Changing them. Tweaking them. Why? Your customers already love them. You're already #1. Is it because you have all these technologists on hand with nothing better to do? You have nothing better than to spend your money and time fixing something that isn't broken?

This is where you can genuinely surge ahead in the field. Spend that time and money on new products. Innovation. Marketing. Anything but messing with something that is already working for you. Most companies indulge in this type of wasteful and largely useless activity. You don't have to be one of them.

There are a few agile principles that sound good but are, in practice, really dangerous. "The best architectures, requirements, and designs emerge from self- organizing teams.". I am almost violently opposed to that entire concept. It assumes too much. What makes you think anyone on a given development team can organize as much as their own sock drawer, let alone a team? There are brilliant technologists out there that have difficulties tying their own shoes.

Architectures are usually applied across many teams, not just one, and they don't magically emerge from untrained or inexperienced people. Also, unless your teams include customers, and most do not, what makes you think a development team, which will contain people that hate to document anything, are going to produce the best requirements? What if you have a team that is primarily noobs? Do you really think they will produce the best designs?

Have any of them been trained on design? Do you know? The best architectures, requirements, and designs emerge from teams that have the right people in the right roles, organized and run by someone who is experienced and in collaboration with other teams that have a stake in the game.

The final principle (about time, too!) is "At regular intervals, the team reflects on how to become more effective, then tunes and adjusts its behavior accordingly.". Everyone who knows me knows that Scrum is not really my favorite way to work. Overall, I just deal with whatever methodology the company chooses to employ, but I strongly dislike some of the largely useless "ceremonies" involved with Scrum. I still remember dealing with the barnyard animal concepts of chickens and pigs and reflecting on how stupid and self-destructive that was. Fortunately, I don't know of any companies that still operate that way.   Many companies, however, do hold scheduled "retrospectives" to review their results and adjust behaviors in in order to become more effective. Except they never do adjust their behaviors. To be fair, I've talked to people in the field that swear by this process.   I'm just not one of them.

You now have a complete picture of what "agile" is all about. Are you truly agile? Probably not. So why don't we redefine what "agile" really means? First, you don't have to do things because those 200 companies over there do things that way. Those companies are not your company. You can define your own agile processes, and those processes can help your company succeed where others, who then become merely pathetic followers, continue to waste their money and their time allegedly conforming to something they'll never conform to. Your process may not be "THE Agile Process". Instead it can be "The (Insert your name here) Company Agile Process".

Another thing to mull over is to create the next Great Thing. Maybe our next Great Thing should incorporate business operations in a world that has changed considerably over the past 24 years. Just remember, T-Rex was agile in his day too.

Clinging to something that doesn't work for your company and will never work for your company is not the way to go. Rather than addressing and solving your problems, it reinforces and contributes to bad behaviors and lost revenue. You're just laying a generic happy word across the messy chaos that is your actual existence and calling it handled. It's so much easier to do it that way. Maybe so. It's also much more expensive.

Strategize. Innovate. Create a methodology that works for you. Move yourself and your company forward and let the rest of the world catch up to you.

CHAPTER EIGHT
I LIKE BIG BUGS (AND I CANNOT LIE)

I expect this chapter to be highly cathartic. For me, not for you. It's all about testing. After 40+ years of spending much of my time attempting to educate people who don't want to be educated about testing, I get this one last shot to try to beat some understanding into our beloved, but determinedly ignorant brethren. You really can't lead IT efforts without understanding the nature of the work that goes into it. What is different this time, for me, is that I don't owe anything to anyone. This isn't a tell-all book – no names are named, but if you recognize yourself or your company here, well, shame on you. Put aside your ego and horrified self-denial and learn something. Then course-correct and go change the world armed with truth instead of wishful thinking. To get all biblical on you, the truth will set you free. Free to go in other directions and explore new paths forward.

The very nature of testing (to say nothing of logic) tells you testing takes place once something is done. You can't test something that doesn't exist. You can explore the idea of that something and discuss pitfalls and benefits, you can design initial tests, you can find issues in design, but you can't actually test something until you have something to test. The underlying truth here is that TESTING TAKES PLACE AT THE END OF SOME CREATIVE PROCESS WHEN SOMETHING IS AVAILABLE TO EXAMINE.

You can, and should, test a piece of the whole earlier in the process. "Shift left" testing is pushing the testing function as early in the process as possible. In IT, we typically call this unit testing. Doing this type of testing will find errors with that piece specifically, but it will not give you good information about how that piece will perform with all the other pieces in place, in an appropriate environment, at the appropriate time. In other words, it is a poor indicator of the customer experience.

Enter "shift right" testing. This is testing the system from a customer perspective in an environment crafted to be as production-like as possible. All testing is good and can find errors, but if you do this type of testing without unit testing first, you find more errors (bugs) at the tail end of the entire development process, which in turn impacts production dates. Both types of testing are valuable and need to be done.

Performance testing is normally done last, because it is testing load against your overall infrastructure and you don't want to halt it because something basic doesn't work. Say one of the functions of your application is to take customer information and save it. If it has bugs (errors) and can't take customer information or save it, you can't verify what happens when 1,000 customers do those things concurrently.

This brings us to a basic, universal truth about testing and quality overall. Look, really look, at the above. If you have no robust testing processes now, you cannot lay those testing processes on top of what you have and expect it to take less or the same amount of time to produce your products. It takes less time to produce crappy products because you cut corners in order to either make a specific date or make more money. Or both. This is the point at which you, as a company (or company leadership), need to decide what "quality" is going to mean to your organization specifically. It is not going to be cheaper than what you do right now. It is not going to take less time.

Ultimately, it can be a differentiator in the field. It's very hard to build a loyal customer base with crappy products. Some companies have made plenty of money that way, but eventually they paid the price in lost market share.

Let's look at some highly visible quality examples – I'll try to do that without naming names.

There is a huge company out there that used to have a stranglehold on the IT field. The entire field had to accept and pay for their updates because they were the only game in town. If you didn't upgrade, your software would not work with anyone else's software and the company selling the software would not maintain the old versions or solve any problems with the old versions. Everyone had to update as updates were available and they paid plenty for that privilege.

Their code was so buggy that almost every company I worked with waited for Patch 2 or 3 before installing it, in hopes it wouldn't bring their entire system down. Their customer support was looked upon as a joke – your problem went on a list of thousands of problems. It would get fixed, or not, sometime. Maybe.

As terrible as this seems, let's look at this from a corporate point of view. Specifically, their corporate point of view. The company made much more money producing (something) on a given date and collecting all that dough than the millions it cost to fix problems after the fact. Think about that. In other words, that model worked for them, at least in the short term.

Everyone, universally, hated doing business with them. As soon as viable competitors were available, they lost a huge amount of market share. They're still Big, but they are far from what they were back in the day. Their founders and core team members are millionaires and billionaires now. So were they wrong? Or were they right? That's your call, not mine. But regardless, they didn't play the long game.

They continue to be a player; I recently read about their CEO talking about producing software that benefitted mankind as a whole and how much their people do for charity and the region in which they're located. A very far cry from "take this software and bite it", isn't it?

Frankly, it bothered me they had the percentages associated with their people that donated to charity; what an individual does with their money is none of their business and I've known companies that try to force that issue in ways that do not resonate with their employees.

Regardless, companies exist to make money and proliferate themselves – a company is an "it", not a person, and although I admire and respect any human being that gives to those less fortunate, I'm not really drinking any of that koolaid regarding a company known for a relatively ruthless disregard for their customers having an epiphany and suddenly developing empathy and a heart. Then again, I'm a QA professional. We're wired to be skeptical.

Now go to the opposite end of the spectrum. If you already have competition and cannot afford to spend millions of dollars fixing your code in production, you cannot afford to operate the same way. Also, some products are genuinely time-dependent and some are not.

You need to set quality guidelines that make sense, financially, for your particular company. If you have competition and produce buggy products, your competition will use your failures to steal your customer base (both current and potential), and in this dialed-in world, your customers will publicly dog you and influence others to take their business elsewhere as well. The smaller you are and the more competition you have, the more important quality becomes to your bottom line, simply because you can't afford to make any big mistakes.

If you choose time as your #1 consideration, be aware this policy works best when you have a captive audience that has no choice but to use your products, and it will backfire on you when other options present themselves. There are times, however, when it's worth it from a corporate perspective. Choose wisely.

Time may also trump quality if you produce something with definite and distinct due dates attributable to the nature of your products. Something like software that assists U.S. customers with filing taxes, for example. That software would have to incorporate the latest tax laws and be available in time for the opening of the tax season. Or perhaps a game or retail support product that is required to be ready for the holiday buying season. Some companies have to position their product for launch at a specific large conference in their niche. These are examples of "hard" due dates. Failure to meet them damages the company in some substantial financial manner.

Many companies choose to use Scrum, which is an agile methodology, for their software development. Scrum is time-based. In most cases, the block of time ("sprint") is set at two weeks. Requirements are broken down into workable pieces and placed on tickets through some sort of management tool (often Jira), reviewed by the team, estimated, and a number are pulled in to be "worked" for the next sprint. This work includes development and testing; the final product(s) should be working and ready to move to Production. If the work produced is just part of a larger whole, these finished pieces might be hidden from the customer until everything else is completed, which might be a number of sprints down the road.

This is quite the challenge from a testing perspective. Why? Because the pieces you've completed cannot really be tested until they can be verified in concert with the other pieces. So you may be moving pieces of code (and hiding them) in Production that will not work when the other pieces are available. The only testing that really gives you an accurate picture of your user experience takes place when all of the pieces are complete, are tested in an environment that closely resembles your Production environment, and tests can be executed in a logical sequence that mirrors how the applications are actually accessed and used by your customers.

Regardless of whether you use Scrum, if your due dates are set to every two weeks, you will likely be working with a scenario where all hell breaks loose every two weeks until the team is beaten up enough to not be particularly motivated by due dates and they simply move all uncompleted work to the next sprint. If your dates are arbitrary, no problem. If, however, you have hard dates that are set by the nature of your product or a goal that must be met for the financial stability of your company, you cannot afford to let this kind of burnout and ennui take hold.

Short, arbitrary dates lead to all kinds of bad behaviors, particularly in regards to testing and verification.

First, in most shops, tickets are pulled into a sprint and development staff and test staff then begin their work. If this is how you operate, you're already behind. A ticket should not be "ready to work" until it has tests associated with it. Why? Do you want your test bank automated? If testing staff do not start test analysis until tickets are pulled in to work, tests will not be available until development is complete and automation will be done after the product is in Production.

Your test automation will be roughly a month behind, and tools that are particularly useful for unit testing and "test left" activities will instead be used only for regression testing purposes (stuff that was working still works), assuming that someone notifies the automation test staff when tests are changed or updated.

Next, development is going to be late. Development is always late. Code might be made available for testing a day before it is due to be moved to Production. Sometimes it arrives even later. This means the QA staff are working nights, weekends, and holidays every two weeks trying to "save" the team and get things tested in time for Production. The big problem with testing is, well, testing is specifically designed to find errors. That is their job and they're quite good at it. So you now have a bunch of errors to address at the tail end of a two-week process. The entire team is now involved in the madness.

During this time, all kinds of bad behaviors will surface.

[103]

QA staff might be told to skip tests that less informed and experienced team members feel are "unimportant". This makes testing staff crazy, although one could argue you have to be a few bricks shy of a load to get into QA in the first place. Your best testers are going to ignore you and do whatever they have to do in order to test what they know they need to test. Overall, you should not be telling your testing staff what to test. They should be telling you. There's a reason QA staff tend to find so many errors and it's not because others know their job so much better than they do.

QA staff might be told to close errors (or someone else on the team will close errors) and they will be told "they're never going to be fixed". This is one of the worst things you can do to testing staff. You are training them to not do their jobs. If they follow those instructions, they cease to be useful as a quality professional. They will stop writing up errors they think the team won't fix. They do not present an honest and complete assessment of the code under test to the team. You want to de-motivate your quality professionals and make them look elsewhere for work? Show them their work doesn't matter by closing everything they find as unimportant. Unless you're in a zero-defect shop, no one expects every error to be fixed immediately, but if an error is an error and is not fixed, it should remain in the backlog.

Testers might be told to not write up problems – to just contact the developer and it will get fixed. This puts the tester, the team, and potentially the company at risk. There is no trail showing testing was done, errors were found, and errors were fixed prior to Production. Failures in Production will be attributed to the tester. This is actually a correct attribution. They did not do their job and allowed themselves to be seduced into non-professional behavior. They've become a quality ho'.

This can cause a company to fail an audit. At the very least, the company can lose both time and money trying to sort out what happened, why, and who did what to whom so it doesn't happen again. Don't do this to your testing staff. If you are part of that testing staff, write up every anomaly you find. It ultimately protects you, your team, and your company.

With short timeframes, it is extraordinarily difficult to do unit testing (individual testing of each ticket or piece of code), a task that has traditionally devolved to the developer, functional testing, which is testing of that piece of code in context with the rest of the code, and regression testing – ensuring code that was previously working is still working when the new elements are added. Performance testing may also be part of the mix.

What often happens is unit testing is not done – the developer does not have time to do it. It is possible this task will go to the tester, but this is actually a poor use of their time, as testing of individual units involves creating mocks (fake input) and takes place in environments that do not adequately mimic a production environment. Errors will be found, of course, but testing will have to be repeated and expanded when all of the work slated to move to Production is complete.

The remainder of the testing may be abbreviated according to risk. Skipping tests always involves risk. It is up to the company to determine what level of risk is acceptable. Often this is done by assessing whether a failure with an individual test would result in holding up a move to Production. Generally speaking, QA staff do not like to skip tests; they are risk-averse by nature.

You want your QA staff this way; their proclivity towards making damn sure everything is working correctly protects both the team and your company.

I find it particularly pathetic that team lateness is still, after all these years of working with testing as part of the development process, blamed on testing staff. I wish I had a quarter for every time I've heard "Testing takes too long.". Mama needs a new pair of shoes and a Bugatti to match. Really? We're still hearing that shit after 40 years?

The actual culprit is likely your project methodology and poor estimation/expectations. No one can wave a magic wand and make 3 days' worth of tests take one day. Development is almost always late but you rarely hear "Development takes too long. Design takes too long.".

Sufficient time to complete the test/fix/retest cycle is never planned. Code is often not unit tested, which means errors found at the tail end of the process will be more numerous, resulting in multiple fix/re-test cycles. Tests may be skipped, resulting in errors in Production. None of this is a fault of QA.

I've been addressing these issues with polite political correctness for a very long time. I'm over it. All of your testing staff are over it too. It's time to wake up and get a clue. Is the field going to go another 40 years flailing around in ignorance? If something is late, it's the team. The whole damn team. If it happens all the time, your current methodologies and processes are not working for you.

I have suggestions for streamlining the testing process; you'll find them in the Quality Ownership chapter. There are a few things worth mentioning here, however.

While AI cannot, at least at this time, take the place of a decent test analyst, it can be better at determining and documenting unit tests than your development staff, primarily because your development staff do not have either the time or the interest to become good at it. It can establish your unit test base and create the kind of test cases that make it easy for your development staff (or SDETs) to automate.

You can also experiment with having AI do the automation. Overall, test automation of unit testing early in the process makes automation efforts in general more useful and with a better ROI. It can help feed and make your CI/CD efforts more successful and will cut down on the number of errors found during functional and regression testing.

Consider starting your AI journey in the testing space with unit testing. It's a gentle, non-threatening way to accustom your technologists to the benefits of AI, as handling unit testing is something none of them want to do, and anyone who automates tests will actually appreciate test cases stepped out with the amount of detail that makes their job easier and more accurate. It also adds value while giving the AI space time to become better at other aspects of testing where it is currently failing to meet its promise.

If you're going to get serious about software QA, you're going to have to invest in software QA. That means looking at your bottom line and determining what quality products and processes are worth to you. Let's talk about cost and time.

If you are willing to innovate, particularly if it involves saving money and time, take a look at the chapter on Quality Ownership. Otherwise, I'm going to assume you want to go the traditional route. That means you're an Agile shop that claims it follows agile methodologies.

How do you estimate what you're going to need and how much time it will take to become adept at Software QA? Here's a simplistic example:

1. You have 10 agile development teams of roughly 4-5 people each.
2. You want automation. Bad.
3. You want performance testing.
4. You'd like to use AI.
5. You don't want to worry about this; you want to hire someone else to worry about it.
6. At the moment, you have a couple of people you've pulled from other jobs and called them "Quality Analysts". There are no testing tools. You end up dealing with Production errors with every release and it's impacting your business.
7. Your applications support smart phones and PCs. You have no statement as to what OS versions you support. The general expectation is "everything".

Here's what you're going to need:

In year one, you'll need one QA Director, 2 QA Managers (one for automation and one for functional), 10 Quality Analysts, a test management tool, and some sort of device farm tool like Browserstack.

During this year, your test base will be built, your automation framework will be built, your test environments will be built, everyone will be trained, and your development staff will adjust to having QA staff on their teams.

In year two, depending on the size of your applications, you will need either 10 automators or 5 automators and an AI tool. It is likely you will need to make the investment toward the end of the year.

During this year, your test base will be become relatively solid, your automators will automate smoke ("sanity") tests, and your AI tool (if you've opted to go that route) will be installed and functioning. Your engineering staff will adjust to having automated test banks available and unit tests defined.

In year three, you will need a performance test specialist and associated tools.

During this year, your entire software QA function will be complete, your functional test base will be solid, automation of your regression test base will be underway, AI will be performing QA tasks for unit testing, performance test processes will be put in place, and your engineering staff will adjust to the additional services and handling errors that come up from testing infrastructures.

There are no specific expenditures in year 4, but you'll be "done" in this year. That's right – if you have pretty much nothing, it will take about 4 years to build something meaningful, functioning, and useful. There's absolutely no doubt in my mind that someone else will come by and tell you how much less time and less money it will cost, but remember rhinestone promises. Anyone can tell you anything. It will take you at least 4 years, and that's if you hire great staff.

Costs for the above? I can only give you rough estimates based on the U.S. at this time. Year 1 will cost roughly 1.5 million. Year 2 expenses will grow to 2.5 million (staff and tools from

Year 1 plus new staff/tools). Year 3 will grow to 3 million and this will be the QA budget going forward, unless your development staff grows.

Are you hyperventilating? No one said quality is cheap. If you choose not to invest, that's on you. You can cut corners and choose not to invest what you need to do the job – when you're still a mess after five or ten years and have lost money screwing around, maybe you'll rethink your position. Are there ways to cut costs? I'm going to say not if you want optimal results.

You can do without managers if you hire talented leads instead, but you'll need more of them and it will expand your costs. In the above example, a director cannot handle 21 direct reports effectively. They'd be spending 3 days per week just in 1/1 meetings. Every direct report in your organization should have one hour of "face" time with their manager. If you do not hire enough people to do the work, your actual expenses and timeframes will expand accordingly. 2 people cannot do the work of 4 people.

Are there ways to cut timeframes? Yes. Those vary so much according to your applications and operations, however, it's too variable to address in any useful way here.

This is just a very simplistic example; every company is different, the complexity of their environments and applications differ, their politics differ, and so on. This was just offered to give you an idea of how much it costs to be a real cowboy. Think of it this way – if you're going to ride the range, you need a good horse….

What you can expect during those 4 years is immediate positive impact starting in year one and improvement in every year thereafter. Even adding one trained, competent analyst to your staff will improve (something). The biggest problem with doing this gradually is that often you end up not fully staffing. Ever. It expands your overall timeframe considerably and you'll be deceiving yourself and others, because it's likely you're going to tell everyone you have QA when in reality you have targeted, half-assed QA on limited parts of your product line. That might be the best you can do at this time – if so, you need to be prepared with communication plans for any fallout. It's infinitely better to just rip the bandaid off and man up if it's possible financially.

Let's move on and take a peek into the admittedly twisted psyche of a tester. I love to test. I've always loved to test. I like breaking things. I've received a pile of awards for finding bugs. The more heinous the error, the more I enjoy it. Those of your staff wired the same way are going to laugh when they find something that will give the rest of the team nightmares. I still remember and enjoy particularly destructive bugs I or one of my staff found years ago.

Some of my favorites? One of my staff put a live link into a field on a screen for a given customer that resulted in every report for every customer in the entire system showing a picture of Mr. Spock at the top saying "Live Well and Prosper.".

There was a developer getting snotty with a QA staff member who asked about character limits and made the mistake of saying "there are no limits – it depends on your equipment.". The tester pasted the Declaration of Independence into a name field. It brought the application down for days.

I once worked for a company that had a devops team make a last-minute tweak in Production. Most testers are highly resistant to testing anything in Production and this was really no exception. When the beleaguered tester accessed our system through Facebook in Production, every customer had his picture on their profile. I tried not to laugh. I really did. I laughed until I cried. When I found out customers were calling to find out why a middle-aged white man was on their profile, I had to disconnect until I could control myself. I have to give props to the Engineering VP on that one. He "got" it. He put the tester's picture on his internal profile for a week. Ah. Good times...

You might say the first two are "edge" cases. An edge case is uncommon, extreme, or unconventional. You might say that no normal person would ever do those things. Well, comparatively speaking, no one in engineering is especially "normal". The problem is bugs that can cause your application to become unusable are, at the very least, a serious risk for your company and need to be fixed. They exploit a weakness that can impact your customer base. You cannot assume every user of your system is altruistic. Some may want to injure you. Some are merely curious as to what will happen.

If you make the right hires, your testing staff will be creative, evil people that have been lied to so often they don't take anyone's word for anything. They raise skepticism to an art form and will want to test everything. If you're going to have these people reporting to someone, that someone has to be woven from the same cloth so they can hire the right talent, understand their problems, and celebrate them when they excel, which will include those times when they find something really vile.

Finding good automation staff is easier. They are builders, just like your development staff, and they just want to code and (usually) be left alone. Again, an experienced manager is important here – there are common problems that come up with automation and you need someone who recognizes them and helps the team solve them immediately. Not by getting in and critiquing code, which is micromanagement and unnecessary if you've hired the right people, but by reviewing results and providing some mentoring.

Speaking of mentoring, it's highly likely your QA staff are going to need to be trained in order to really become something special. College degrees are great on the automation side, since SDETs and automators need to be able to code and they can learn those basics in school. Test analysis and functional testing, however, at least in the United States, is not taught in college, except as a by-product of the development process. You need to be trained and you need to practice in order to become really good at test analysis.

There are several schools of thought on test analysis and as much as it will make others scream (since most people belong to a camp), either one can be effective at training testers how to look at something and figure out what needs to be tested.

One view (standard), for example, might teach boundary testing (no characters, minimum characters, maximum characters, beyond the maximum characters, special characters, numerics, alphas, spaces, emojis). I like teaching that way because it's more succinct and easier for me, personally, to teach to a newbie.

The other (exploratory) might make use of heuristics (rule of thumb). That method will use more verbiage. Like

"Overwhelm the field with characters.". As long as the many ways of deconstructing code are covered, it doesn't matter what it's called, but everyone on your test team should have working knowledge of the basics and understand each other.

The purpose of training is to introduce these ideas to the tester and expand their understanding of what to examine during the testing process. If you have a good tester before they're trained, you'll have a great one after they're trained and practice.

There are a variety of ways to get this training (look up software test training on-line), but expect to invest in it. Even if you have a senior resource, if they've never been trained, they need to be trained. You can also have some fun and have your highest-ranking QA person run a bootcamp.

By the way, automators should also be trained in test analysis. They may never actually do much in the way of test analysis, but it's a big plus for them to understand the process and their functional compadres and as they progress in their careers, they may have to manage functional testers. That's hard to do if you don't understand or appreciate their work.

Every QA person on your staff should either have a mentor or be a mentor. Becoming a testing guru involves a lot of learning and no one is born with all that knowledge. You grow into it.

So this is my final thought regarding testing from the 10,000-mile view. Nothing in QA matters more than practical experience. When you look at a resume from someone who has hopped around every few years, it's really not the fact that they've hopped around. In today's landscape, they might have been laid off a bunch of times and lay-offs are "no-fault" situations and do not reflect on the individual. However, you cannot have had 10 years of experience at 5 companies and expect to come on board as a senior or lead. At least not with me. Years of experience do not equate to position. Why? Because every year or two you were starting over as a newbie. You haven't worked anywhere long enough to pick up the skills (both hard and soft) required to step into more senior roles. If you're a QA person, think about that and try to aim for positions where you feel you can stick around long enough to learn/exercise the type of leadership skills you need to progress.

In closing, I'm going to take a phrase from Monty Python and possibly taunt you a second time. No, I'm not going to tell you your mother was a hamster and your father smelt of elderberries. It's just this. QA people are trained to tell the truth. Eat your hearts out. If you're a fearless, quirky person (with or without the outrageous French accent) with a strong sense of right and wrong, to say nothing of unbridled curiosity, it might be just right for you.

CHAPTER NINE
AUTOMATION MUST DIE

What?!? Automation must die?? What kind of bizarre, neanderthal mindset is that?

I started out in development. I've learned around 18 languages – some proprietary, some not. I've worked on creating an automated test language. I did nothing but automated testing for over 10 years. I've larned me a few things and would like to share them with you.

First, programmers grow up and become directors, VPs, CTOs, and sometimes even CEOs. When you say "automation" to a techie, you're speaking their language. It's what they've done, what they do, what they believe in, and what they understand. Even non-technical upper management staff love the sound of "automation". It is equated with cost savings, speed, and efficiency.

Those of you that think test automation is a relatively recent obsession are wrong. Every company, everywhere, has wanted test automation since testing became a "thing". At least 50 years.

Here's the sad part of that history. After all that time, we still don't get it right. There is little to no understanding of what automation does well and what it does poorly. To say nothing of how much time it takes and how much it costs. To add to the pain, there's a lot you need to set up from a process perspective to make it work successfully.

Automation, as it exists today, must die.

It is a bottomless pit into which you throw money and resources, with very little benefit in return. Companies exist to make money and proliferate themselves. It stands to reason they need to invest in things that give them a return. Instead, they often buy into stupidity and hype because it sounds good. Anyone can tell you anything. Smart people and smart companies are going to look at the results and decide if continuing in the same direction makes sense. Simply put, companies cannot afford to waste money because of empty buzzwords that promise a lot and deliver nothing of value. We need to rethink automation.

Let's talk about why automation efforts fail. Some companies, in fact many of them, believe SDETs (Software Development Engineers in Test) are like Quality or Test Analysts on steroids. They can design test plans, manually test if needed and automate. So they want to hire only SDETs. The problem here is SDETs and Quality Analysts have different skill sets, interests, and abilities.

Yes, there are a (very) few exceptions, but overall SDETs are developers. They look at your systems in the same way your developers do. They miss the same tests your developers miss. Most would prefer to have their toenails ripped out with a pair of pliers than manually test. They want to code. And they want to code from stepped-out test cases because that's easiest and fastest for them.

The majority of technologists do not understand or think like your end users. What is also both charming and naïve is that they do not consider improper use or input into your applications. Quality Analysts are a tad more evil. And realistic. Hiring solely SDET staff is not going to get you even a decent test base, let alone something comprehensive. Consider that it doesn't really matter how fast or who runs a test if it is the wrong test, run at the wrong time, in the wrong environment.

You are still going to require a Quality or Test Analyst to define your test base – or at the very least someone who understands and is capable of thinking like a potential customer and has the training and discipline to turn that knowledge into tests.

What's more, even your lowest level SDET is going to cost you at least 25% than a senior Quality Analyst. A senior SDET will likely be 40% higher. Or more. These are highly compensated employees that need to be understood and deployed intelligently.

Let's consider where you are going to find the bulk of your errors. In my experience (and yes, yours may be different), roughly 97% of your errors will lurk in new stuff. 3% will reside in existing code negatively impacted by the new stuff. Unfortunately, testing existing code cannot be skipped (regression testing) because often that 3% is critical to the business. So both new and existing code needs to be tested.

This can be a problem in terms of the value of automation and its contribution to your development efforts.

Typically, your SDET staff are going to want stepped-out test cases with verified results in order to do their work. This is the easiest and most efficient, convenient way for them to do their jobs quickly. Unfortunately, creating stepped-out test cases is the long pole in the tent in terms of Quality Analysis. In fact it takes so long it can, and often is, created after the fact.

[119]

This means your SDET staff don't begin automation until the test effort is complete and the code is in production. Hopefully their automated tests become useful for future regression testing, which contains roughly 3% of the errors you want to catch. In other words, your expensive automation staff are contributing to future regression efforts only, and even that can be of very limited use in typical workflows, which tend to lack both structure and process.

Do you really want to deploy your most expensive staff and most time-consuming processes on 3% of your problems?

Agile processes are rarely very structured. In fact, the development process in many companies is somewhat of a free-for-all, but maybe not quite as much fun. If you use Scrum methodology, you might have some standard "ceremonies", but often these are incomplete, meaningless meetings held just so you can say they're held and in reality are completely side-swiped or abandoned when things go awry. And things always go awry.

Let's examine what kind of structure, discipline, and process you would need to ensure automation success as it exists now.

When a Quality or Test Analyst creates a stepped-out test case, they need to know when that test gets automated so they don't run it manually. That means some field on the test case itself should indicate automation status.

As code comes through and gets modified, the existing test case base has to be examined and updated, re-submitting it to the SDET staff. If this is not done, the automated test will fail and the SDET will have to determine why and whether the new results are expected or an error. In other words, they spend time doing analysis that has already be done – which is frustrating and time-consuming activity that slows them down. In addition, if their tests are built to be dependent, one test failure can cause an entire test set to fail, again wasting precious time and money.

There is rarely a one-to-one correlation between a manual test and the associated automated test. This means the SDET staff need to take care to comment their automated tests and test runs to ensure they can easily find and update them as tests are updated by the Quality/Test Analysts.

Engineering teams rarely consider the time, cost, and effort involved in maintaining an automated test base. If your automated test base is not maintained, it will likely be out of date within three months and completely unusable in 6 months.

This is not true of manual tests. Why? Because manual tests do not do byte-by-byte compares and kick out anything that doesn't match. Even if you change technologies and move things around on a page, if your application (for example) requires a name and address, it is likely any future iterations of that page will also require a name and address. Every automated test you have, however, will fail.

So beginning your automation journey is predicated by having something to automate. If you have no manual test base to start with, Step 1 will be creating one.

There are other peculiarities with implementing automation that are common and need to be understood. Your SDETs are technologists and will need to create test data. This is something automation does extremely well and it can be utilized for manual testing and performance testing as well. More on that later. If, however, your environment is shared with manual testers or others, it is important automators do not use any data used by others, as their tests will change the state (and therefore the usefulness) of that data.

In addition, technologists often use garbage as test data. Your Quality Analysts, for example, might create a name like "James O'Donnell". Your technologists will create a name like "Aaaaa B'bbbbbbb".

There are both advantages and disadvantages to this. The disadvantage is your test environment will end up populated full of garbage, particularly if the automated tests don't clean up after themselves. In other words, everything created is later deleted as part of the test or test data used is returned to its original state. A test environment full of crap makes it difficult for anyone other than technologists to work in and much of the data is likely to be old, corrupted stuff no one can use. On the plus side, if things go awry, the nature of the test data shows in an instant if the problem was generated through a manual or an automated test. Figuring out which automated test caused the problem can be quite the challenge, however, as the test data generated is largely random.

Once you have designed and built the automated tests, they are commonly strung together into a test run to be executed together. Problems often occur at this juncture as well. They are a normal part of test automation and are usually easy to fix by your automation staff.

First, as a human being works through your applications, the result of some actions will result in a system response. It might be an order or customer number, for example. It doesn't matter from a system perspective whether the source of the input is human or generated from an automated test. In addition, actions may be "stamped" with the date and time. These are all areas that have to be handled programmatically by the automation tests. They can either be ignored ("masked"), where the automated test results (compares) ignore mismatched data in those fields, or stored as variables so whatever that value is can be retrieved and used later in the tests or test run.

With apologies to all of the talented technologists out there, let me offer the most simplistic example I can.
I am creating an order and when I press the big red "Buy" button, the system responds with an order number. I can use that number to track the status of my order.

An automated test can capture that system response and store it as a variable – such as A$. Then later when a test requires the order number, the input will be A$, which will retrieve the system response (the order number) captured earlier.

So often the first run or two of automated tests will be locating and programmatically handling system responses and time stamps.

Another common problem is that regardless of the source of the input, it takes your applications X amount of time to respond. If, for example, it takes 9 seconds manually, it will take 9 seconds automatically too. Automated tests often (I was tempted say "always") run faster than your applications can reasonably process them. That means your automated tests may have run perfectly individually, but when you string them together into a test run, the whole thing barfs (technical QA term) rather spectacularly.

This requires the automator to go back and add waits to their code. Generally, there are two types of waits. Implicit ("slow everything down") and explicit ("when I perform this action, do not execute the next test command until a response is received"). Explicit waits are more efficient than implicit waits.

At the same time, your automator may insert timers, particularly if you have no other performance testing going on. While comparatively crude, well-placed timers can tell you certain transactions took X time and compare that to how long it took the last time the test was run. This can give you a heads-up on potential performance issues.

If you've stuck with me thus far, you now have a handle on basic automation testing and associated problems. Daunting, isn't it? There is, however, a reason (or two or three) to invest in test automation.

The primary reason, and the one that makes executive staff salivate, is that it's FAST. The second is it will run the same tests reliably, repeatedly, accurately, and quickly without any expensive human interaction other than to kick off a job (which can also be automated) or review results (which cannot be automated). It is not easy or fast to get to that point, however.

So let's talk about where test automation really shines and where it is less effective.

Automation is great for producing large quantities of test data quickly. This can enable a variety of testing efforts, including functional (manual) and performance testing. To be useful for functional testing, however, technologists have to resist nonsensical field entries that are easy to generate but hard to use, manipulate, and debug from a human being's perspective. In other words, Antoinette Smith is easier for a human to work with than Abcdefghij Klmno.

If you make use of smoke tests, that is a great place to start automation efforts and can yield a relatively fast return on investment and be viewed positively and with relief by your engineering staff. A smoke test is a very basic set of tests, using valid data only, to verify business-critical functionality is working as expected. QA staff are often asked to run these sort of "sanity" tests (they normally take a human between 15 minutes and an hour) for a variety of reasons. If not automated, a human being might need to be available, for example, to run smoke tests after a devops tweak at 2 AM. An automated smoke test, with very little training, can be run and validated by anyone on the engineering team at any time.

Another area where automation is far superior to human testing is for unit testing, particularly when verifying the attributes of an element on a screen. What are attributes? Things like "I am enabled, I use this font and font size, I am this color, I start on this byte and end on this byte".

Human beings can check these things, but it's a royal pain in the... well, it's a pain. Often these are things you cannot see with your naked eye, or areas where human beings might have problems (like shades of color). To be honest, often these tests, even if they fail, are not enough to halt code promotion to production, but they still need to be checked and what is a time-waster and time-consuming for a human being is both fast and easy automatically.

In addition, unit tests are usually just testing just one chunk of a larger feature. In and of themselves they may be pretty meaningless from a customer perspective. They are often tested in a "lower" environment (not matching production), using "mocks".

A mock is test data that is not real. Sometimes you cannot create "real" data because the code used to create that data doesn't exist yet. Sometimes mocks are used because the test data needs to be very specific (like a vendor with more than 50 orders per day) and it is infinitely easier to mock the data than to hunt for real data that meets the requirements. The use of mocks is common, normal day-to-day business for development and unit test efforts

Yes, it does cause problems down the line when "real" data is put through the code. Mocks do not normally contain invalid input, where real data can be quite a mishmash of stuff, and sometimes mocks are incomplete, since the developer or unit tester is just looking at one specific element or part of a page.

Automation is great for creating test data and mocks, and testing just a piece of a piece of an intended feature is not a problem. The real issue with this is that automators are unlikely to get stepped-out test cases or tests with known, valid results to automate until well after a test effort, or sprint, is concluded. So where they can really shine and benefit the team is often not even an option for them.

Automating a smoke test or maintaining an automated regression base is not part of the responsibility for any one given team's automator during their development cycles and would hold them up from producing their normal work. Often automating a smoke test takes place in an automator's "spare time", of which they have none, as most are assigned to teams and are perpetually behind because they don't receive their work until after code is moved into production.

I believe it's time to mention automation vs functional (manual) testing. Automation is not good at functional testing. I realize many technologists out there are going to be howling about how wrong I am, but remember I did that work for over 10 years. It's not really a question of whether automation CAN perform a functional test, but rather whether it should, and whether the end results are worth the money and time it takes to produce something that is reasonable. Do I believe in automation testing? Of course I do. But I believe it needs to be employed for those things at which it excels, and functional testing is not one of those things.

When you are running an automation test set, you are taking the results of the current test run and comparing them against the results of those same tests run earlier. Any discrepancies, even if something is off by a pixel, is going to kick out as an error.

You have to tell an automated test what data to capture and compare, unless you are doing something akin to a "capture/playback", where everything is captured and compared. Capture/playback is a nightmare in terms of automated test maintenance and most automators will yak up a hairball if you even suggest such an outdated, heinous test technique.

If the output of an enter on a page 1 is a page 2, for example, the automated test will look for and capture that you've landed on page 2 and call it a day. What this means is that automated tests are not observant. Human beings are observant. We're also curious by nature. That means if page 2 looks funky for some reason, we, as humans, are going to notice it and poke at it. Automated tests do not notice anomalies that are just off somehow and investigate them.

Can you build automation tests to perform a functional test? Yes. Is it worth the time and effort? If we're talking about smoke tests, yes. For the rest of your test base? Maybe not. It depends on the size of your test base and how/when you want to automate your regression test base. It's not something that can be done by personnel assigned to a team producing work every X number of weeks.

It has to be handled outside normal project work, and time, personnel, and expense will be required.   It will have to be rigorously maintained to remain useful. This means you will likely require an independent team of SDETs (whether on-site or off-shore) that specifically create and maintain your automated regression test base.

It also means your structure and processes had better be disciplined enough to make that possible.   To be quite frank, I don't think most companies have that level of discipline. We can all piss and moan about it, or we can deal with reality, look at what we've got, and figure out a better way. Automation, as it exists right now, needs to die.

Based on what I'm seeing right now, I think it more likely AI will replace some, if not all, of the work of an SDET. In addition, I think it will save a company time and money to marry AI with test automation practices in order to push the ability to have automated test sets run earlier in the development processes as unit tests. This "shift left" type of thinking would enable AI and automation to focus on the 97% error rate in the new code, rather than the 3% in the old code.

Automating a regression test suite would still be a thing, since regression testing tends to take a long time (comparatively), but these tests could be handed off to automators/AI after the sprint.

Does this mean automation and the need for SDETs will go away? No. It means, however, their jobs will morph. SDETs are technologists and AI is a tool. It's a tool they need to become familiar with and exploit to their greatest advantage. If they don't take this opportunity to learn and grow, instead of investigating and recommending solutions, solutions will be imposed upon them.

IT grows and changes at a tremendous rate. You must keep up. This is a perfect time, while the technology is still pretty new and not great yet, to get in there and figure out how you can use it to benefit your company and your own skill set.

New things can be scary, and a bunch of idiots telling you you're going to be replaced by a "thing" isn't helping matters at all. You do not have to be terrorized, however. You have to be proactive. It's the unknown that tends to be scary. So make it known. Go out there and learn about it, study it, and use it – make yourself an expert. Do you realize how valuable that would make you? You're a technologist and this is a new set of tools. This is what you do and what you know. Go for it!

CHAPTER TEN
TESTS OF DOOM

This book wouldn't be complete without looking at Performance Testing, A.K.A. The Tests of Doom.

I've done performance testing using JMeter and Loadrunner, but what I've done and what a performance specialist with decent tools can do is like me telling a master baker I once made a mud pie.

Performance testing falls right into Rhinestone Quality. Everybody wants it. Pretty much everyone says they have it.

Not.

Performance testing involves a number of interesting things. It can tell you how your system performs under load, can assess a variety of stresses during a given timeframe or under certain conditions, and can predict future bottlenecks.

The reason so many companies give lip service to performance testing is that good performance testing is expensive. It involves expensive people, time, and tools. Most companies cobble something together using open-source tools or have their SDETs (Software Development Engineers in Test) use some intelligent timers and place strategic implicit waits in their automated code and call it done.

If you have less than five thousand concurrent users at any given peak time, that might be a good strategy for you. The results won't be as detailed or instructive, but that might be enough to keep you afloat during prime time for your system or app. As long as you're willing to take a hit to isolate, debug, and fix problems in terms of time, that's not really a great solution, but it's a solution that might work for you at this stage in your corporate growth.

For now, however, let's abandon that rhinestone reality and stop-gap measure to talk about the Real Deal.

Look, it's your company – not mine – and you need to take a look at your bottom line and decide how much performance testing is worth to you. If your applications went belly-up under peak load, how much would you lose? That's really the only number that matters.

At the time of this writing, if the loss would be less than 250K, you aren't really ready for anything especially sophisticated right now. So going the route of bastardizing some open-source tools and making any automated tests capable of informing you of glaring problems should work for you until your size calls for something better. Don't bother to hire a performance specialist, as they're probably too expensive for you at this stage of your growth, you'd just frustrate them, would fight purchasing them the tools they need, and overall working for you would do nothing for their resume or career. Try pushing your SDETs to stretch their wings and learn new things, which is good for their careers.

If, however, you stand to lose more than 250K – per failure – it's time to Get Real and protect your company against preventable, catastrophic loss.

Speaking as a hiring manager, your first challenge (and it's a big one) is finding a performance test specialist. They are the most difficult roles to fill in the QA space. Although there are a few exceptions, they are also going to be the most expensive. Their value is such that they are rarely looking for a job.

You might wonder why performance specialists are at such a premium. First, learning anything in the QA space is not done through education in the traditional sense. Your degrees and certifications may at least indicate someone open to learning new things, which is great, but it is not indicative of either talent or experience. The only way to learn QA is through learning it on the job and through what is actually an informal type of apprenticeship. So learning performance testing is going to be a combination of personal interest, learning, and experimentation coupled with absorbing some of the "tricks of the trade" from a talented peer or mentor. It's very difficult to find a specialist, period, let alone one that is willing (or has time) to show you the ropes.

Next, performance experts specialize in a very particular niche and need to be comfortable with and understand aspects of a system/application that other members of the team do not. They need to be able to automate tests, like an SDET or developer, although the automation required to do performance testing is comparatively simple. However, on top of that they need to understand your system architecture, be relatively expert in devops, databases, and any specialized transmission protocols, to say nothing of a healthy amount of understanding of the various quirks of a large number of devices.

This means a performance expert has to be a weird combination of architect, DBA, SDET, data analyst, and more. This puts performance specialists at a premium and once a company nabs one, they normally do not let them go. Unless you are a ginormous company that has a reputation for attracting top talent, it can take you a year or two to find a performance specialist. Yes, it's that bad.

There are things you can do in the meantime as you ready yourself for that kind of hire. These can save you and your company time, money, and effort if you prep in advance. These are the things your performance specialist is going to ask for or need when they come on board. They're going to be pretty impressed with you if you're ready for them.

Try gathering non-functional requirements and statistics in advance. Non-functional requirements are things like what your peak load is, when it occurs, and how common transactions are spread across that load. What do you consider good in terms of response time for those transactions? What devices do you support and what is the spread of those devices?

If you don't have this information, you'll have a bunch of largely meaningless guesses and a big pile of ...nothing at the end of your performance testing challenges. I've included a list below of common statistics/non-functional requirements, but I warn you it is not exhaustive. I'm not a performance specialist. I'm a groupie and a fan. These should, however, give you an idea of what your performance guru is going to ask you for and help you launch your performance test efforts successfully.

1. System/architecture diagrams
2. List of devops and monitoring tools currently in use
3. List of development languages/tools currently in use
4. When are peak times?
5. What is your highest peak load – number of concurrent users, number of transactions?

6. What is your anticipated peak load?
7. What is the balance of transactions going through during peak load (such as 20% login, 40% search, etc.)
8. What are your expectations in regard to response times? What is it now and what is desired? At what point is it unacceptable?
9. What is the balance of devices accessing your systems at any given time (such as Android, Apple, PC, etc.)
10. What is the load on 3rd party applications at peak time? What are your expectations in regards to ensuring your partners can handle your anticipated loads?
11. What tools do you have in place right now for monitoring system health?

So you've found your performance test expert and gathered the non-functional requirements you need to jump start the process. What's next?

You're going to need an exclusive environment dedicated solely to performance testing. It needs to be designed and provisioned to mimic production as closely as possible. Otherwise, you might get some useful results, but you will always be extrapolating what *might* happen based on the restrictions of the test environment, rather than what will actually happen. Furthermore, the environment will need to bypass any security you have in place to prevent bombardment of data from hostile entities.

This is going to be an expensive and relatively time-consuming process, but it is critically important. You want accurate, informative, "real" results.

Why does the environment need to be exclusive? Here's the fun part of performance testing and the reason they are The Tests of Doom. Good performance testing is destructive. Not occasionally destructive or kinda destructive. It is testing annihilation and cataclysm. Other teams cannot share the environment as it can and will be brought to its knees if done correctly. This makes it unsuitable for use by other teams.

Good performance testing does not just verify your apps will perform acceptably under peak load. It will tell you at what point your applications start to degrade and when they'll die a terrible death. This is of huge benefit to your company as a predictive tool.

Say your peak load is 5000 users. Performance testing might tell you your new features don't impact your system under a normal load. Maybe it will also verify it executes flawlessly under peak load. Your performance specialist is going to continue to add load and rerun the tests until they find out when your response times start to degrade. And continue to add load until they determine at what point your product becomes unusable.

Say that point is at 7600 users. This is outstanding information for you and your company. You now know you can go to production as things stand right now, but that work needs to be done as your user base grows towards 7600 users. It allows you to put that development work into the queue according to your growth plan, rather than in a panicked rush once the need becomes critical.

What is even cooler about performance testing is that it can tell you exactly where the transactions are bogging down and depending on the tools you're using, exactly what programs are causing the issues, thus saving your staff and company time and money trying to determine root cause of the degradation of services.

That brings me to the next piece in the puzzle. Tools. Everyone has their biases and favorites. If you are an owner or executive, you'll likely be biased towards open-source tools – A.K.A. "free". While I get that and feel open-source tools are great choices for many things, in this particular case you'll likely get what you pay for.

Performance testing is not an area in which to skimp. My advice, if you're going to go through the pain and expense to hire an expert and build them an environment, is that you actually trust them to do their job and recommend the best tools for the job. I don't know your company and their set-up. I've used lots of tools. At the time of this writing, I like Blazemeter and SmartBear. I especially like them because of their understanding and support of the devops landscape. That said, however, tools need to support and be tailored for your specific environment. There are a ton of tools out there and a good specialist will recommend tools specifically geared for optimizing tests in your particular environment. Listen to them and try not to wince. Performance tools are expensive, but you do not hire an expert and then refuse to give them the tools they need to do the job for you. Or maybe you do. It's certainly common enough.

Don't be that company. Look at your bottom line, the financial impact of failure, and either belly up to the bar or go home.

Either investing in performance testing or not investing – yet – can be the right answer. But be honest with yourself and recognize the likely fallout from failing to actually invest in performance testing. If you go with the rhinestone version where you say you have what you clearly do not, you need to have some sort of communication plan in place to explain any catastrophic failures in production. That might or might not work, but unless you're the owner, you're putting your company, yourself and potentially your teams at risk.

Performance testing, done right, is a piece of art. If you don't have it right now, it will change your world, enable your growth, pinpoint exactly where your weaknesses are, and allow you to charge forward with your mission while confident of success.

CHAPTER ELEVEN
AI IS LYING TO YOU

AI (Artificial Intelligence) is one of the hottest topics in the IT world right now. I'm primarily going to talk about AI as it relates to software testing
.

There are two rather distinct vocal camps regarding AI. One hates it. The other loves it. Proponents in both camps are savagely trying to discredit and annihilate the other. As in all things, the answers and the truth likely fall somewhere in-between.

AI is new technology. It was created by technologists. And (wow!) are they excited, for good reason, about what AI can do.

But it doesn't do it right now. Because AI proponents are mostly trying to sell you their AI products or ideas, you wouldn't know it because of all the buzz, but AI is not ready for Prime Time just yet. If they get smart, start listening to their detractors and tighten up what's wrong, it might get there, but for now, they just attack their critics and do their best to discredit them personally.

I read one blog post where the AI proponent disdainfully pointed out their opponent drove an old Subaru. Um, really? You used one of the most regulated fields in the world to make your point? And you criticized someone who drives a car recognized as one of the most reliable and safe vehicles in the world? In the words of Bugs Bunny, "What a maroon!". And no, I don't drive a Subaru. But it makes sense to me that someone who cares about quality would own one. And I kinda like the look of the Crosstrek Wilderness...

This is just one example – the pushback from the other side is just as condescending, arrogant, and obnoxious. Ego plays a huge part it. Overall, when someone is trying to sell you on something because it affects their bottom line, including ideas, you can expect bloodshed. It's pretty entertaining to watch if you aren't in the middle of it.

Both sides are using pretty dirty tactics to get your attention and move you (and your company's money) into their camp. On one side, they're telling you the technology sucks rocks and your jobs are going to be replaced by something vastly inferior. On the other side, they're telling you everyone is already doing it, you're behind, and if you don't embrace it and join their camp, you're going to be history. In both cases, they put your future on the line.

Both sides are full of shit. I think what is most pathetic about the whole situation is that if those two sides actually got together to build something useful, they'd change the world. But that would require them to squash those ginormous egos, listen to and respond to feedback, and work together. I can't see it happening. At least not at that highly visible level.

So what is the truth? Hey, it's a relatively new technology and I don't have all the answers. I just know what I'm hearing isn't it. The pieces don't add up. That's what QA people do. We look for anomalies and report. So I'll share some of my observations and make a few recommendations. And yes, I've taken the time to educate myself. You should too. I've taken classes, read, attended lectures and presentations, experimented, looked at results – the whole ball of wax. And it fascinates me.

What automation does exceptionally well is crunch numbers. What AI does exceptionally well is crunch data. It siphons in huge amounts of data from sources it has been trained to examine and presents answers to questions based on putting those results together in a format you've requested and therefore find useful. It can spot redundancies. It can detect patterns you, as an individual, can't see because it might take you years to assimilate and process that huge amount of data. It can do in seconds some things that might take you years. There is tremendous and unbelievable potential there for huge strides in medicine and science.

But.

AI needs to be "trained". At this time, a human being has to guide it as to what to look for, gather, and interpret, and how to present that information. That training is being done by technologists. Then human beings have to examine that data and determine if the results have validity and applicability.

I read an article (naturally from a company that was a proponent of AI and selling AI products) that said most people believe the results of AI. Now THAT, my friends, is scary. Because AI is lying to you. Not deliberately. It is doing its job, canvassing the information available and presenting it to you. AI is ARTIFICIAL intelligence for a reason. The source of the information can be questionable and the information itself can be bogus. AI is not intelligent in the same way as a human being and does not consider the source before presenting information to you. It was trained on where and how to look for data and it presents it as gospel. The less information or valid information available from its sources on a given topic, the more ridiculous the results. It also reflects the limitations and prejudices of its trainers.

Regardless, you need to be aware that some information readily available and accurate can be bypassed and false information presented to you, because the AI has not been trained correctly, nor has it been trained to properly discount information lacking certain attributes.

Let's use some real-world examples. There are tons of examples about AI failures to accurately present valid information or results from the opponents of AI, but I, personally, am not an opponent. Merely a skeptic – it has to do with my QA training. I am by nature skeptical. I question everything, I test, and I make up my own mind. If you're in QA, you likely do the same.

You can have some fun with this yourself. I was at a conference where a group presented a list of tasks/questions it gave AI and the bizarre, incorrect results they received. Don't believe it? I do. Test it.

At the time of this writing, Google is presenting all of us with an opportunity to play. Their AI overviews state they are experimental. Given the content of some of their results, it's a good thing they've covered themselves. Type in a question. Any question. Take a look at the AI overview results. I don't know what your results might be, but mine were terrible. And incorrect.

What did I ask?

I asked how many carbs were in an oreo cookie. The answer was 21 grams. There are not 21 grams of carbs in an oreo cookie. Information regarding the carbs in an oreo cookie are available all over the place. On their website. On the back of their packages. On various engines that monitor nutritional data. Yet the answer was incorrect.

The AI either canvassed the incorrect sites for this information, or it did not "understand" that "an oreo cookie" means one cookie. Regardless, consider the potential impact of this incorrect information. First, if I have to restrict my carbs, I am not going to eat a single oreo cookie. Next, I am not even going to BUY oreo cookies. And this is where it gets serious. The AI overview, if I believed it, would result in a lack of a sale for Mondelez (Nabisco). Not a big impact for that giant – one bag of cookies, right? Except that information will be presented to anyone that inquires. So the incorrect AI results can potentially negatively impact sales. What does that mean? Given that companies exist to grow and proliferate themselves, my thought would be lawsuits. When I see this kind of example, I predict some serious legal messiness in our futures.

Then, because I'm a glutton for punishment, I asked how many companies were using AI for software testing.
I knew better. Seriously, I knew better. But I did it anyway.
The results I received on that last week said over 75%. The results I received this week said 44%. 75% or 44% of what? Why did the numbers change? To be completely honest, both results made me laugh. But if you believed the results, either time, you might be traumatized.

The fact is that there isn't enough information available yet to give any kind of meaningful results. If the AI were properly trained, it would not be canvassing purveyors of AI services for

[143]

an answer to that question and if there were no answers available quantified with numbers (of X number of companies, Y% use AI for software testing), then the AI should report that statistics are not yet available on that topic. Last week's results were pulled directly from a blurb from a company selling AI products/services. So that means if the data is pulled from a company or information that is a tad skewed (OK, it can be a total pack of lies), then those are the results you're going to get. Those results are not accurate. They're wishful thinking from a single company trying to sell something.

So what kind of conclusions can we make from this? Is AI a pile of crap? No, it is not. At least not across the board. It appears to me that the efficacy and accuracy of AI is dependent largely on several things. One, it has to be properly trained to understand the question or task. Second, it has to be properly trained to pull data from the proper sources. And last, but certainly not least, it has to be properly trained to remove data that is questionable because it does not meet the basic criteria for an unbiased, scientific result.

Let's take that further. In every case, the problem is not what AI CAN do. The problems are what it DOES do. And all of that revolves around how it has been trained. So let's consider that for a bit.

When I started learning about AI, I was both fascinated and repelled. Attracted and horrified. AI was created and is being trained by technologists. Have you worked with technologists? I have – for most of my life. They are incredibly excited about AI. And AI has amazing potential, so I totally get it. I started my career in systems analysis, have learned 18 languages, worked as a DBA, and done all kinds of things that aren't necessarily what you'd consider "QA", although I learned all of those things to support some kind of QA initiative. I understand and appreciate technologists. The thrill is definitely there with AI. But while I get it and have to say that in some ways I share it, there is also an element of horror.

Technologists are excited about what AI can do. What they haven't considered is how it should and should not be used from an ethical standpoint. There are no rules around use of the technology. Just as it can be used for noble purposes, it can also be used for terrible purposes. Consider that many companies are purchasing the technology to fulfill their mission, which is to grow and proliferate themselves. In other words, marketing and convincing you to buy their products and services. To distill that down, they seek to influence and manipulate you. Lots of people seek the same. Politicians. Social media influencers. Once AI is fully adopted, you may never read anything not tainted by self-interest again. It may take you ten times longer to get reliable answers to simple questions.

To illustrate, consider the simple act of searching on your phone or PC. You may type in a very specific phrase or term. What you get in return is a huge list of ....stuff.... that does not contain the specifics in your search. If I look for a specific company, for example, I expect results where my search term is respected and returns information for that specific company. But it doesn't. You can pay to have your competing company appear at the top. Is that ethical? I don't think so.

From a customer perspective, it is beyond irritating. If I enter a generic search, I'm fine with companies paying to appear at the top of the list. But if I'm specific, I want specific results. I have abandoned many searches because it sucks up WAY too much of my time paging through the results looking for the specific thing I asked for in the first place. There's an assumption that if you're interested in one type of product, you're going to be interested in every single similar product on the market. That is not, however, how human beings operate. Sometimes we are looking for something very specific and sometimes we are browsing. Search technology was not really originally designed to do these irritating or unethical things. It was purchased and modified to work this way in order to make some companies more money.

Do you pay any attention to the ads that pop up when you're searching? Those are the result of what you've been interested in at some other time. What is poorly understood, again, is that human beings don't really work that way. We have curiosity. We may look something up one time to assuage that curiosity. We do not want every single associated reference to that thing shoved in our faces for the rest of our lives.

When we're shopping, often we look, we purchase, and then we're done. Again, we don't want similar items shoved at us forever. It's irritating. When we purchase something, the site will send us emails (sometimes multiple times per day), forever. All of this takes up our time and effort.

Now multiply that irritation by a thousand. Is that what you want? Does any of that improve our lives in any way? Does it provide any benefits for humanity as a whole?

It becomes even worse. Technologists are training the AI. Actually, you can just say "human beings are training the AI". Human beings have prejudices, limited experience, blind spots, and they are training the AI.

Technologists are particularly blind to the ethics surrounding their beloved technology. They also tend to be blind as to how normal people operate and use their technology. In some ways it's a type of innocence, as they themselves would not use the technology unethically. Unfortunately, their technology can (and is) sold to others who don't have any type of moral compass. Or their technology is imposed on others that would just as soon pass.

Overall, the wrong people are training AI. AI is being trained to lie to you, or at the very least present limited, prejudiced, misleading, or flat-out inaccurate results. And it's an artificial intelligence that is being trained by people that don't understand or cater to customers that are human beings.

That said, let's move on to AI as it pertains to software quality assurance. Software quality assurance isn't understood very well in the field, in spite of being around as a separate field for over 50 years. Much of that is due to the organizations being run and managed by staff that "grew up" as software developers. It's difficult for builders to understand destroyers – their focus is different and the value they provide to a company is fundamentally at opposite ends of a spectrum. The AI is being built and trained by technologists – who are builders.

In order to understand the problems with that, you have to have some understanding of a typical software development lifecycle. Here's one basic take on it:

1. Someone (usually a product owner or BA) with an understanding of what is wanted in terms of the product from a customer perspective develops specifications of some sort telling the software development team what to build.

[147]

2. Those specifications, which are often in a visual format designed by UX designers, are broken down into pieces of workable chunks which are put on "tickets" to be worked through by the software team. This is often done by the product owner or BA as well, although the software team may add technical tickets to support the work to be done or break them down even further to make them easier to work.
3. The software team meets to decide what tickets need to be worked within their "sprint". A sprint is a largely arbitrary chunk of time put aside to develop product functionality. Some companies produce (something) weekly, every two weeks is most common, and some have 4-week or 6-week sprints. It depends largely on what the company produces, how long it typically takes them to produce something meaningful, and how often their customer base can tolerate change.
4. Once the team has decided what to work on, developers start designing code and QA starts to define the test base. From a QA perspective, this task is often started late as they are still testing and working on code produced from the sprint before.
5. As pieces of functionality are completed, someone (this should be the developer, but rarely is due to time constraints) tests that piece. This is "unit testing".
6. Once all of the pieces that are going to be presented to the end user/customer are completed, QA tests that it works as expected, doesn't break, and that none of the existing system(s) that should still work do work. This is "functional testing".

7. QA will find errors, write them up ("defects" or "bugs"), and the team decides what should or should not be fixed before that functionality moves to production. Normally the team will either fix or defer any functionality not ready for production.

8. QA is also responsible for writing up test cases – a step-by-step guide to the test with expected results. These are handed over to someone (usually a Software Development Engineer in Test – SDET) to automate. Normally automation takes place or starts after the functionality has moved to production. This allows that functionality to be retested at a later date automatically.

This is a very basic, typical scenario. Let's talk about the most painful and time-consuming problems with this cycle
.

Specifications take a long time to put together, and breaking them into workable tickets also takes time.

Development teams in general have a tendency to be extremely optimistic about how much they can complete during a sprint and they do not allow any time for the fix/retest part of the process. Often a technical complexity or two throws a wrench into the works and developers end up having to work overtime and push hard to get what they agreed to do actually done in the time period allotted for their task.

Agile project methodology is actually pretty good at hiding how long it actually takes to complete new functionality or features. If it doesn't make it on time, the team moves that particular ticket or set of tickets to the next sprint. So the time it takes to develop, unit test, fix, and submit code for QA testing is a bottleneck. It is common for code to be submitted the day it is due or even later – forcing QA to work OT, the entire team to work OT, and for hard decisions to be made in terms of what functionality to move to production. It is also common for development to skip unit testing or push that to QA staff who are already pressed for time.

Logically speaking, everyone either forgets or doesn't think about testing. The key to good testing is running the right test, at the right time, in the right environment. Who or what identifies the test and does those things is immaterial. All testing is good and can identify some problems. But the tests that are most meaningful to the company and influence their market are the tests that ensure the product does what is supposed to do and doesn't break from a customer perspective. That means all the pieces have to be present, all of the tests identified/run, and the environment has to closely mimic what an end user will actually be using.

The tests you run at the tail end of the process are more meaningful than those run early in the process, even though those early tests can cut down on the number of errors found later.

And QA staff are good at their jobs. They find errors. They find errors AT THE TAIL END OF THE PROCESS, throwing the whole team off their game.

Finding errors and reporting is their job and they do it. While writing a test outline (or "plan") is a relatively fast process, it starts too late in the process – at the same time development starts. That means developers don't gain the advantage of knowing in advance what tests their code needs to pass. As time is at such a premium, it is almost impossible for a Quality or Test Analyst to write stepped-out test cases for those tests. This in turn means SDET staff do not get test cases in a format easy for them to automate in time to make any appreciable difference to the mission of the team.

QA is often still testing a given sprint when the rest of the team has moved on to the next sprint. It is common for QA to be two weeks behind the rest of the team, so the vicious cycle continues.

So where are the opportunities for AI to make a difference in this cycle? For some ideas as to morphing automation efforts, see Chapter 9 ("Automation Must Die"). To sum up that chapter here, I believe AI is a better way to go for unit testing.

In terms of next steps for testing, I'm going to throw something out there on my personal wish list. I wish AI proponents had (go figure) actually asked quality experts about what they wish they didn't have to do and what takes them the most time to complete. We'd be further ahead.

First, AI is bad at defining functional test sets. You can cry about it, tell me I'm full of shit, and fling poo in all directions if you like, but reality is reality. AI does not, AT THIS TIME, understand human beings and its "understanding" is further hampered due to poor training and direction from technologists.

In addition, some data needed to properly understand and use a new application is not available yet and therefore cannot be gathered and be made into a pattern that can be applied to similar things. That limitation is recognized and being worked on, but AI does not have imagination. It has the following of a similar pattern that might look like imagination, but it is generally not imaginative or innovative. It is limited by the data available on those things it has been trained to examine. That said, I do not want AI defining test sets for functional testing, which is customer-based, end-to-end, and includes invalid data of a type that requires imagination and/or innovation.

You know what I could totally get into and support? Something that would take a structured test outline ("test plan") and write stepped-out test scripts for me. And no, I don't want to have to write my outlines in pseudo-code. Human beings are going to have to use those outlines (and stepped out test scripts) and they are not going to want to learn pseudo-code. Writing test scripts is BORING and takes forever. The fun part of the process is determining what to test and the actual testing itself, where I can watch something go belly-up. I love that part of the work. Most testers love that part of the work.

All I can say, AI puppeteers, is "chop, chop". Go forth and improve my life. And don't tell me AI can do that already. Maybe it "can", but it doesn't. Go make it happen and (geesh), partner with real testers, OK? Not your technologist compadres.

If we could improve and cut down on the time required to create the kind of documentation that makes tests easy to recreate (by anyone!), automate and/or feed into other AI processes, life would be good.

While we're on the topic, it makes sense to me that if you want AI to become the Next Greatest Thing, instead of shoving something potentially scary down someone's throat, you'd partner with them instead. ASK. Ask them what part of their jobs they hate the most and would rather not have to do. Then focus the attention of AI services on those things specifically.

Rather than ten bazillion people pushing back at you, you'd have ten bazillion people screaming for your tools. Talk about missed opportunities.

To top it off, AI providers are advertising and pushing things they really don't do very well (yet). So don't do those things. Yet. Push what you know and can prove AI does better. Push what you know anyone who tries or tests what you have to offer can't disprove as bogus. Right now, it's way too easy for your detractors.

Partner with your intended customer base – actual human beings doing the work. Address their pain points and stop trying to make something not human into something uniquely human. AI fails at those things. It can gather data about what humans do and copy patterns, but it doesn't really understand the "why" behind anything. It is not imaginative. Or creative. At best, it can mimic those things. My own thought is why even bother with that shit? People do it better. So focus on things people either do poorly or can't do particularly well. Then work on the other human-like services in the background until they're better and ready for prime time.

I've thrown down the gauntlet. I double-dog-dare you to pick it up...

CHAPTER TWELVE
IT'S ONLY A FLESH WOUND...

I'm going to specifically talk about building a career in Software QA here, but much of it is applicable to any field. If you've already built your career, you can read this just for fun.

First, if you've chosen QA as your career, congratulations – you've chosen one of the most difficult aspects of IT in which to craft a career. Building your career is going to involve working through some pain, picking yourself up, and continuing to Fight the Good Fight. On the plus side, there are few careers in the entire business world where you're paid to tell the truth and this is one of them.

If you tell someone what they want to hear instead of what is true, failures can occur that impact your customers and the company bottom line. You need to recognize up front that the nature of your work is going to be at odds with many people and acting more or less as a corporate conscience isn't necessarily going to help you move up the ladder. Particularly when the ladder is already strewn with people that don't share your sense of right and wrong or believe honesty is the best policy.

Let's start with your qualifications. There is a decided prejudice at this time for hiring college graduates with STEM (Science, Technology, Engineering, and Mathematics) degrees. When looking for your first job, having such a degree will help you out. However, the more experienced the hiring manager, the less your degrees and certifications are going to matter. They're going to want on-the-job experience. That doesn't mean degrees and certifications aren't good – if everything else is equal in a job search, they can help tilt things your way, but they aren't as much of a differentiator as in other fields.

QA is one of the few fields in IT where your degrees and certifications aren't going to help you be better at your job. If you don't have a STEM degree or other meaningful certifications (which only matter to inexperienced hiring managers), you still have options. Look for large companies with a large number of staff. These companies are more likely to hire newbies and offer the training and mentorship you'll need to become successful. Specifically, look for the word "associate" in the title. If you've done related work – like working in customer service where you had to learn to write bug tickets, work with development or devops, and solve customer problems, bring that up. If you've worked in project or product management (attempting to herd cats), bring that up, as well as your abilities regarding anything touching understanding and managing systems operations.

In QA the only way to learn the job is to do the job under the tutelage of a mentor. I wish apprenticeships were more of an option – that concept is perfect for software QA.

QA people need to be trained. They aren't just magically born. Most "fall into" QA – they were involved in something else, somehow got involved with a testing effort, found some bugs, and (poof!) were moved to QA. Without training, however, no one can really move forward easily and their progress (and therefore careers) are slower to mature. It takes training, time, and practice to be a good tester.

There are a few ways to get at least basic training without mentorship. None of them include a college degree. It really depends on where you work (or want to work) and what you want to do with your future. Some certifications are more valuable to you than others. If, for example, you want to work internationally, at the time of this writing ISTQB certification is recognized internationally and it might help you get the kind of international opportunities you're looking for. If you want to learn your craft, i.e. testing, there are many on-line opportunities and certifications available.

The basic competency in software QA is that you know how to test. If you've never been trained, you don't. That doesn't mean you haven't tested some things and found some valuable bugs. It means you've missed other things because you've lacked the training and discipline needed to discover and think about those things. So learn how to test. Otherwise, you'll miss too much to be useful.

What about SDETs (Software Development Engineers in Test)? Isn't a degree useful in those cases? Yes. SDETs need to know how to code, but the nature of that coding and purpose is different than that of a Software Development Engineer. If an SDET is incapable of test analysis and is not trained in how to craft automation frameworks and automated tests, they will be limited in the value they can provide to a company.

So get trained. If your company does not provide mentorship or training, invest in yourself and go get it yourself on your own dime and your own time. Good companies that want highly trained, valuable employees are going to invest in you if you have potential, but they are not required to build your career for you. Your career is your own responsibility. They are paying you to provide X services for Y pay. As long as they do that, they are meeting their side of the bargain.

You are in charge of yourself and your progress. Don't allow anyone or anything to hold you back. Although we all want to believe others will recognize our potential or reward us for our achievements, hard work, and dedication, often that is simply not true. We need to take charge of our own destinies and move forward. It is so much easier to blame others than to blame ourselves. Inertia is as strong a force as gravity. Ultimately your career is not your boss's or company's responsibility. It is yours.

You'll be a happier person if you are realistic in terms of what you expect. In Software QA, the usual progression is:

QA Analyst
Senior QA Analyst
Lead QA Analyst
Manager of QA
Director of QA
VP of Quality
-or-
SDET
Senior SDET
Lead SDET
Manager of QA
Director of QA
VP of Quality

Notice the titles merge at the management level. We'll talk about that a bit later. To begin with, someone becomes a QA Analyst or an SDET. Some large companies will make a career path more standardized by having QA Analyst I, II, and III. Likewise with SDETs. This is to provide an easier and more understandable career ladder. The expectation is generally that you will spend 2 or 3 years at each level.

What does that tell you? It tells you that a senior resource has at least 6 years of experience. Based on what I've seen, maybe it's as little as 5 or as many as 8, but the 5-year mark is a good general rule of thumb.

Most people I've known want to move into management, even if they shouldn't really work with other human beings. What you need to understand, however, is that you may spend your entire career at a senior level. There is absolutely nothing wrong with that – in many fields worldwide being a senior (whatever) is a hard-won honor, pays well, and deserves respect. Regardless of what you want right now, it's fine to change your mind later on. Human beings do not need to remain static. Our dreams can morph into new dreams as we move on.

If you are expecting (at the time of this writing) a six-figure salary and are not at a senior level or above, your expectations are unrealistic in the United States. I won't say it has never happened, but I can say with confidence it rarely happens. I can't give guidelines as to what to expect, because it depends on where you live. Do some research and find out what is common in your region and don't ask for something out of line with that. In either direction. I've met (and interviewed!) some people who vastly underrate their worth, and some who vastly overrate their value in the market. Ask for and expect something realistic.

Here are some interesting things about interviews. First, most of us are somewhat nervous during an interview. That's OK. Keep in mind you are interviewing each other. Be prepared by researching the company and asking questions. It might help your nerves to understand the hiring team WANTS you to be the one. They do not want to interview 2,000 people. They want you to be the perfect fit so they can be done with this process and get back to work. They will have to keep any appointments they have already set up, but if you're the one, they'll cancel anything out further than a week and stop looking.

I'd like to also encourage you to not ask certain questions, at least not in IT. If you ask me about work/life balance, you're telling me you're an inexperienced noob and that my company will not be able to make you happy. I'd really rather you get your first total disillusionment from someone else. I don't know if that's true in other fields, but it's certainly true in IT.

Secondly, don't ask only questions that have nothing to do with anyone but you. That means if you ask about benefits, pay, work/life balance, etc. and don't ask one question about the company and how it operates, don't be surprised if you're passed over for someone who shows more of an interest in the company.

You do not have to discuss/negotiate pay or benefits until an offer is actually made. It depends on the nature of the position, but personally, what I want and expect at my level of experience is going to differ depending on the opportunity. I'm going to require a certain level of pay, period, and will not budge below that. However, what I actually expect and request will differ according to the opportunity itself. Is it building something from scratch at a small company? Or is it managing multiple world-wide teams and functions for a huge company?

Again, determine what is equitable on your own, after you've done your own research, talked to the company, and have a clearer picture of what they're looking for and then stick to it.

By the way, you do not have to disclose what you made at your last company. Actually, you don't have to disclose what you've made at any company. You should, however, be prepared to ask for what you're looking for from their company.

I've turned down a ton of jobs. In fact, I think I've turned down more VP jobs than anyone I know. Title means nothing to me. It wasn't always that way – I learned that through painful experience. Titles shouldn't mean anything to you either. Titles are free. Budget, authority, respect, pay/benefits, and trust all matter to me. I've had management positions that were more in line with a VP, director positions that were really lead positions, and pretty much everything in-between.

That said, if you are offered an opportunity that does not meet one of your core requirements, like pay, be prepared to walk away. I've never regretted a single opportunity I've turned down. But I've regretted a few where I let my ego win out over my common sense. Again, TITLES ARE FREE. It doesn't matter if I'm called Queen of the Known Universe if I make minimum wage.

If you don't have a good feeling about the opportunity – for whatever reason – trust your instincts and walk away. I realize you might think that's easier to say than to do, but remember I have actually done that many times. It's infinitely easier to gently and politely pass something by than to spend the next X number of years in a situation you don't like and really didn't like from the get-go. I don't expect a company to bother talking to me if they can't afford me, but if money is the issue and I walk, the company, now realizing I have a bottom line in terms of pay, may come back and re-negotiate. Regardless, set your standards, which should be based on good research and your personal needs, and stick to them.

There are a lot of different people in this world. It's hard for me to relate to someone who just wants a paycheck, to be comfortable, and has no particular ambition. I'm not like that myself. For the purposes of this chapter, I'm going to assume you want to be the best of the best, be wildly successful, and that you have a healthy dose of ambition in your makeup.

Whenever you take a job, make your mind up that you are going do everything you can to make that job and that company successful. You are going to view it as the opportunity it is, get everything you can out of it, give everything you can to it, and move on once you hit roadblocks not of your own making and that you cannot change through your own hard work and contributions.

In other words, when you become unhappy – not at one circumstance or event, but pretty much all the time, there's no point or value to you or to the company in remaining. Unhappy people tend to do poor work, it affects their health and personal life, and they cease to promote and contribute their best work and good juju to their companies.

If you're doing your best work and contributing, you should expect to move up the ladder at certain points in your career. If the career path is a I, II, III, Senior, then every two or three years, you should progress up that ladder. During these years, hone your craft. Take testing classes. Volunteer for testing opportunities outside your normal working hours. Read books on your craft. Start following what is happening in your career space on-line.

In all cases, if you do not progress and get promoted as you expect, you have two choices. If you like your job, your coworkers, and your company, ask (even if it makes you uncomfortable) what you need to do to progress. If the answer makes sense, adjust accordingly and do what you need to do in order to move up. This shows your management you are willing to learn and accept mentoring, all of which is good for your career. If, however, it smells strongly like bullshit, trust your nose and start looking for opportunities elsewhere. Bad situations or toxic environments rarely get better. In fact, they often get worse. Save yourself.

Once you become a Senior, your job duties will start to change and the nature of the things you begin to learn and study should change as well. You should be (or should be close to becoming) an expert. You should be learning to mentor or teach others. That means you're going to start to learn how to work with and manage people. This is the point where you start to pick up "soft skills". Soft skills are non-technical skills that are important if you're going to progress towards management roles. Up to this point, your technical skills were more important than your non-technical skills. That is now going to begin to morph.

You may start to lead efforts in those areas in which you are most expert. People will seek you out for advice and help. Either seek training from your company on soft skills (negotiation, dealing with conflict, etc.), or invest in yourself and take those classes outside of work.

It is also important at this juncture to become more than "an expert". You'll want to start positioning yourself to become "THE expert". As a Senior, it's going to be expected your technical skills are top-notch. Make it so. One of the reasons it is important to keep on top of what is happening in your field is so you can remain relevant and be an asset to your company. People who never change, learn, or contribute anything are of lesser value. Oddly enough, it doesn't take long in IT to become a fossil. The field changes too quickly.

As you move up the ladder to a senior position, many weird opportunities may come your way. I strongly advise you to take as many of them as you can. The more you know, the more valuable you become. I've done all kinds of funky things in my career – been a developer, a project manager, a BA, a DBA, technical writer, learned all kinds of devops jobs, and am certified and educated in a crap-ton of things that may or may not even exist any more – or that don't exist yet. All of them, added up, are me. In the words of Walt Whitman "I contain multitudes.". These experiences will make you versatile (which equates to "valuable") and a better tester, a better manager, and a better employee. It also has the added benefit of ensuring you are never bored.

This is also a time to show your support for your team and your company. If you want to be the go-to person, people have to want to come to you. Help everyone you can in a positive way. It might be your team and your peers, not solely your management, that influences the powers-that-be to give you your first management job.

One of the keys to making it easier to get promoted is to make that decision a no-brainer. If you're already recognized for having the skills necessary for the next rung on the ladder, it's much easier for your management to make that happen.

It's at this point, when you're a Senior, that your management (and their management) starts to take a really good look at you and make some decisions in terms of your future with their company. It's important to make your own ambitions known at this point. If you want to move into management, your boss and your company can either help develop you in that direction, or not. Regardless, they need to know you're interested in moving that way. Otherwise, you might remain where you are indefinitely. That might happen anyway, but usually nothing happens unless you express an interest.

The next step up the line is to become a Lead. A Lead can be viewed in two ways. First, it is the highest level "individual contributor" role. Second, it is the lowest level management role. If you've been promoted to a lead position, congratulations! You've either been identified as someone with management potential, or as someone so technically gifted you have the potential to become a principal. Either way, the company doesn't want to lose you. Take this as the compliment and opportunity it is.

Your life as a Lead is going to depend heavily on your boss and your environment. You will also probably be stuck at this level for a while. It might be two to five years, with a strong leaning towards five. An experienced manager – or a good one – is going to take this opportunity to mentor you and teach you the kind of things you'll need when you become a manager. A lesser one will just let you handle the most difficult technical, tactical work without adding to your understanding or handling of managerial responsibilities.

Theoretically, a Lead heads up a group of people, mentors them, removes everyday obstacles to getting work done, and ensures work gets done as expected.

If you don't lead a group of people, your lead position is an indication of technical superiority and not a stepping stone to management. Your next logical step will not be as a manager – it might be as a principal.

If you do lead a group of people, some of the more valuable things you can hope to learn are solving problems, coordinating with other groups, assigning work, mentoring staff, and, if your manager is with the program, handling 1/1s, advocating for your people, recommending training, and advising your manager. I cannot speak for other managers, but something I personally look for at this stage is someone who has an interest in the field and who brings new ideas forward. Your company may also have a leadership program and you should be invited to participate. Your boss may ask you to step in and handle their responsibilities when they are on vacation or unavailable. This is a strong indication of trust and belief in your abilities.

In my opinion, this particular position is one of the most valuable in the company. It's also your opportunity to show everyone what you've got. As an aside, a helpful hint here. Train yourself to almost never say "no". Instead, force yourself to look at everything from an analytical perspective and come up with solutions that can help make the best of bad situations. When you do that, the few occasions where you have no choice but to say no, everyone is really going to listen. Since you're training yourself not to say "no", you're going to learn how to negotiate and compromise with others and add positive mojo to whatever is happening at that time.

That doesn't mean you take it on the chin every time someone asks you for something. It means you don't brush everything off with an automatic "no" without really thinking about how you could make it work. If you choose not to do this, your life will be more difficult. You will end up being forced to do things to which you said "no".

It's the same with learning new things, getting new tools, new processes, and the like. You can either research things like AI and make suggestions yourself, or have people that know nothing about what your group does or how it operates impose those things on you, likely making choices that were not the best decisions for your area. Become proactive, flexible, and collaborate as much as you can.

The most difficult move anyone ever makes is from Lead to Manager. First, there are limited managerial positions available in many companies.   If you work in a small company where opportunities are limited, you might need to move on in order to get where you want to go in terms of your career.

It is very helpful to work for a company that likes to hire and promote internally. Not every company operates that way. Competition starts to get very tough at this stage of the game. There are simply less positions available. By this time, you should have at least 10 years of solid experience. Your competitors will

.

From a corporate perspective, if you hire managers that have not moved through the ranks, there are important lessons they have not learned. There is enough talent available that you needn't make those kinds of bad choices.   When posting managerial jobs, if your job description specifies less than 10 years of experience, you're not really looking for a manager. You're really looking for a senior or possibly a lead to whom you will offer a title. That means you are going to have to invest in a lot of training. Managers in the IT field make over six figures here in the U.S.; you should advertise for and hire the best you can for the money you have available and they need to be qualified for that level of work. If you don't have that kind of money available, hire a good senior or a lead and mentor them to grow as the company grows.

Consider taking your own people and moving those with the talent and potential up - it is far more beneficial to the company than hiring from the outside. They already know your company, your culture, your products, and your people.

The move to management is going to change your life and not necessarily in good ways, so be prepared and be smart. The change in your status should involve a nice chunk of change to help ease the pain. If you make the change for free in order to get a largely worthless title – your mistake. You'll have to move on to another company to get the money that should be associated with this kind of promotion. Still, having the word "manager" in your title will open many doors and as long as you deliberately take the job as a strategic move to get the title and plan to move on, it can work to your benefit. If you pass the opportunity by, you will need to accept that turning down this type of job internally may mean you'll need to move on to move up.

From a corporate perspective, particularly if you're promoting from within, make becoming a manager at your company worthwhile. It sets a good precedent and inspires those in your pipeline to excel.

Let's talk about being a manager. There are a lot of really terrible managers out there. Becoming a manager is "it" for a lot of people. It represents a ton of work, a high degree of expertise, and for many people it's an ego thing that tells the world you're a success.

But.

Managers need to be good with people. If you're not really all that fond of human beings, you're going to be a bad manager. If you don't want to spend your time helping others succeed, you're going to be a bad manager. If you aren't especially protective, don't like teaching, collaboration isn't really your thing, your way of handling confrontation is to ignore problems until they go away (which is pretty much never), you're going to be a bad manager. There is, however, one problem that falls into the Worst of the Worst category. When you become a manager, you must learn to let go. Otherwise, you become a micro-manager. Your life will be miserable and everyone who works for you would rather not.

Becoming a manager is a huge adjustment. You are no longer one of the guys. You are no longer one of the gang. It's just sad to watch managers (and above) that want to hang with the homeboys and talk tech. It's what they know, what they feel comfortable with, and what they've done. Unfortunately, it is no longer what they are.

Becoming a manager changes your relationships. Even if the guys seem to treat you like one of the dudes, they are always conscious of the fact that you now have authority over them; their paychecks, their work, and their environment. You will no longer necessarily be told the truth – at least not all of it. Some things you won't be told at all. People might suck up to you in subtle and not-so-subtle ways. You might find you like that. You might remember, however, that your very best friends are not head-bobbers. People who care about you are going to tell you when you have toilet paper stuck on your shoe.

There are now going to be things that go on you cannot share with them. Those things will affect your interactions and relationships. In addition, you may now have to have some uncomfortable discussions with people who were friends.

Your job has now morphed from that of an individual contributor and you need to let it go. It is likely you were very good at your job – if you weren't, you wouldn't have been promoted. Someone else will be doing that work now. If you were really gifted, that is hard, hard, hard to work through, because most people are normal, not geniuses. The person now doing your work may not be as good at it as you were. They may not do the work the same way you would do it. Take some deep breaths. That is OK.

Your job is now to do the best you can with what you have – not to do someone else's job. You cannot do everyone else's job for them and do your own. Something would have to give. You can do one thing well or many things poorly. Your job is now strategic. Let the day-to-day stuff go – set up and mentor some leads to handle the nitty-gritty of daily work. Then (this is important), let it go. You can tell if things are going well by results. If the results are not what you want or expect, mentor your leads and get it to where it needs to be. Rather than being the best, your job is now to build and maintain the best.

There are working managers, but to be honest, if you have 8 people or more, you won't be doing any tactical work; at least not if you are going to fulfill your role as a manager. If, however, you are a manager of (say) 4 people, you may be more of a lead, and will (and should) be more involved in the day-to-day work than an actual manager. It's up to you to recognize when you've moved to a "true" management situation and when you need to step back and let others handle day-to-day work.

When you become a manager, you become your people. If they suck, you suck by association. The last thing you want to hear is "OMG, Bob's people are AWFUL.". If your people rock, you rock too. "OMG, Bob's people are AWESOME.". What this means is when you are hiring, you need to hire the best people you can afford, and this is no time to get paranoid.

Some managers are intimidated by people that appear to be smarter or better than they are. Again, this is especially true of those who were experts when individual contributors. Losing that expert status and leaping into a pool that might be full of sharks can make what used to be a nice guy into an insecure lunatic. It doesn't have to be scary. Hire people you think are actually better than you were (or are), or that have the potential to surpass you.

The unspoken fear of an insecure manager is that someone who is uber-talented will stab them in the back and take their job. First, if you have a long career, someone you trust will likely stab you in the back. It is, however, much more likely that someone will be a peer or a superior. Not one of your employees. I have to say that in over 40 years, I have never had an employee stick it to me and take over my job. Normally any remaining managers would not want to work with anyone who is blatantly self-serving and poisonous. None of them would trust such a person and you need to operate with some level of trust in order to be successful. So sticking it to your manager is not likely to do any good things for your career.

Unless that manager is so heinous that getting rid of them benefits the rest of the staff, the company, and you're doing it to Save the World as you know it, I'd recommend (whether it's your manager or a co-worker) to never hurt another human being in an attempt to elevate yourself. Karma is a bitch and you're likely just reserving your own place in Hell. That said, it is rarely an employee that is going to lose you your job. It's going to be you, the relationships (or lack thereof) that you've built, or a business decision that has nothing to do with your ability or worth.

So throw away your doubts and insecurities and go for it! Build a team that makes other managers jealous and cry into their Starbucks. That's much more fun.

It's during this time that you will learn about politics. I've already said that QA people are notoriously bad at politics and I am very definitely a QA person, so there is no advice I can give you about how to swim through the murky waters of corporate politics. What I can say is that if you are like me, trained most of your professional career to tell the truth (no matter how painful), if you do things in the best interests of the company, support your peers as best you can, and build awesome teams, you may survive, thrive, and even get promoted. There are going to be a lot of games out there. You don't have to play them, particularly if you're bad at them.

You will, however, need to learn to pick your battles and there will be times you need to be polite, friendly, and helpful to a real asshole you'd really rather choke and leave for dead. What can I say? Do the best you can. Only time and experience will help you learn when you can and can't win a war, and common civility (regardless of your personal feelings) keeps a company moving forward. Ultimately it is you that will decide what kind of person you want to be. Do you want to be paranoid? Destroy your enemies? Is your workplace a battlefield for you?

Does that enervate you? Then go for it. You'll likely need a decent mentor to help you survive. If, however, that isn't you, then navigate around it as best you can and refuse to play.

As you learn to navigate your way through the perils of politics, there are other situations that may come up that negatively impact you or your people. One of the most common is that your company could be sold. There are "sweet spots" in terms of earnings that make companies attractive to investors. When this happens, you'll find your upper management staff strangely unavailable for a period of time and when you do see them, they'll likely appear exhausted. It's against the law for this to be discussed with anyone not involved in the negotiations, so don't expect anyone to clue to you in.

Depending on the buyer, staff will be laid off. It might be a few people and it might be most of the staff. You may get some say in who goes and who stays and you may not. It is not uncommon to have to justify those decisions to executive management. You may be asked for all kinds of process work and statistics.

You are also going to be put in the position of explaining the unexplainable to the staff and conducting individual lay-offs, unless the company chooses to do a mass lay-off with everyone leaving in one room at the same time, at which time the meeting will be run by HR and some executive officers. Once the dust clears, it will be your unenviable task to keep your remaining staff motivated and moving forward.

As you go through all of this and learn, be aware that QA is rarely a field where the members are regarded as team players. That's because when everyone else is being politically correct and nodding their heads, you're going to be asking hard

[173]

questions and telling the truth as you see it. Although that is actually a huge win for any company (the higher up you go, the less likely it is you will be told unvarnished truth), it is rarely viewed that way. If you become too inconvenient, the company will let you go. In addition, if the company needs to lay people off, your people will often be first in line for the chopping block. Although quality is a strong differentiator in the field, testing (whether manual or automated) is not normally regarded as valuable when compared to developers or other staff.

That is just a reality of our field and you need to accept it and move on with grace and whatever tatters of dignity you have left. Help your staff adjust or move on. Wish everyone well. The best revenge, if you need revenge, is to be successful and happy elsewhere.

More learning curves will be dealing with staffing problems, confrontations, performance issues, and budgets. In terms of anything having to do with people, I have some simple advice. Be kind. Even when you have no choice but to let someone go, you don't have to be a jerk. Kindness is free. Try to help them move on.

When you need to talk to someone about a problem, if you've hired the right people, you don't need to beat them up about it. They'll be beating themselves up about it. If you're really angry, you're not in the right place to mentor someone. Invoke the 24-hour rule and wait a day until you've calmed down. People rarely react in positive ways to being yelled at or threatened. Wait until you can address the problem in a way that shows you have their best interests and future growth in mind. Everyone makes mistakes. Everyone. In terms of dealing with your own mistakes, acknowledge them (own it), fix it if you can, learn from it, and move on. If you teach your people to do the same, they will feel safe to admit to, fix, and learn from their mistakes.

Another issue that might be a problem for you is God Complex. This is when you can't admit you don't know something and are too paranoid (or arrogant) to ask for help. The best managers, directors, and VPs I've known are very aware of what they do and do not know and ask for clarification and help. They do not try to tell others how to do their job when they've never done it themselves. There are some people that believe asking for help or relying on others will be regarded as a weakness. It's actually a strength. What's more, it can genuinely motivate whomever you ask, as you are acknowledging their expertise and valuing their advice. Multiply that by 10 if you give them credit for their help publicly.

Once you've been a manager for a while (usually at least 5 years), you can start eyeballing director-level positions. What makes these positions different from managerial levels are the number and type of people you'll have working for you.

Managers usually have staff reporting to them that do the same things. Functional testing or automated testing, for example, but not both. A director, however, will have all of it. Functional testing, automated testing, performance testing, etc. A director's specialty might be automated testing, for example. If, however, they know nothing about functional or performance testing, they will do a poor job in their role. This is one of the reasons it's such a good idea to grab any opportunities you can as you're moving up the ladder to expand your knowledge base and experience.

Director positions are even harder to come by than management positions and there is plenty of competition at this level. It is also common to place a software engineering manager, with no experience as a QA or SDET person, into the role. I'm just going to speak in generalities here, but no one especially enjoys working for someone that doesn't understand what they do. That said, many director positions are political in nature. If that is the case, it's very important to put some managers in place under said director that understand what is going on and who can actually run things.

Director positions are strictly strategic. If you run into job descriptions that say you'll have to "wear many hats", that means they don't have the staff to do what needs to be done and you'll be doing everything, possibly by yourself, for the foreseeable future. Your director title is a rhinestone position. Do you take it? It's up to you. You can look mighty purdy in those glittery boots...

The job of a Director is akin to that of a Manager, with more people doing differing, but complimentary, tasks. Your direct reports and those you mentor will be at the Lead or Manager levels. You will be responsible for everything under your purview. That means if an SDET screws up a run or doesn't meet their obligations, you'll be explaining that and attempting to protect or advocate for them to VPs and possibly C-level executives. If an error pops up in Production, the first person who will be approached (and blamed) is you.

Usually, your home and work life will become one entity. You're pretty much on-call 24X7. If a problem comes up, you're expected to step in and help solve it.

Once you hit this level (and above), unless you are part of the original core team, your position is not going to be particularly safe. It's one of the reasons many Directors and VPs make the Big Bucks. In fact, once you get to VP level, it's likely you'll actually have a contract. Make sure it spells out the conditions under which the company can dispense with your services – that is, make sure you get paid to take a hike. It needs to be enough to support you while you find a new job. At this level, unemployment is unlikely to cut it.

What makes VP positions different from Director positions is that the people who report to them may actually have different skills, abilities, and career focus. A VP might handle Software Engineering, QA, Devops, and Product Management, for example. At this level, it is unrealistic to expect one person to have experience performing the work of every person in every group reporting to them. That's a massive learning curve. It's critical to have experienced people who can collaborate and give good advice reporting to them.

Again, just as in the Director position, a VP is responsible for every problem and every win for every individual in their area. In addition, they will be representing the company in a variety of ways and are expected to be part of the "face" of the company. They will also be expected to represent the company to their employees in a favorable light.

In general, from the position of Manager on up, your job will be handling problems, all day long, every day. It isn't for the faint of heart. The higher you go up the ladder, the more time you will spend in meetings, strategizing on those problems, next steps, and the company's future. Even less of your time will be your own.

[177]

After experiencing all of these things, in my opinion the positions with the least amount of personal time are Leads and VPs. Leads have less time because they are acting at both a highest level of technical expertise and lowest level management role that keeps them hoppin'. A VP is simply responsible for so many people and functions, their days (and nights) are filled with handling issues.

If work/life balance is important to you, you might want to think long and hard about that. It's up to you to set boundaries you can live with, and those boundaries might not make you a good candidate for positions that require a lot of personal commitment. While the money is usually great at a VP level, it's not much of a plus if you never have time to enjoy any of it.

Above all, remember that your career and professional journey is solely your own and there is nothing that dictates to you what has to happen or when. You can change your mind at any time. Until then, fling yourself into whatever resonates with you and persevere. There are people out there who have much less intelligence or intention than you have that have been successful. You can surpass them, whether that is in pay, position, or happiness.

CHAPTER THIRTEEN
THE NUMBERS GAME

Lucky Chapter 13 is all about metrics. How appropriate. Metrics are misused, misunderstood, miserable for many, and just in general, all of the "mis" words. There is no aspect of business, including QA, that cannot be influenced (positively or negatively) by numbers.

Most practitioners in the IT field don't merely dislike metrics, they despise metrics. This is because they've been on the receiving end of the misuse of metrics on an individual level.

Metrics used to measure individuals can make your place of business into a sweatshop focused on all the wrong things. Attempting to use numbers to measure the value of individual human beings to your business with no additional context may, on the surface, seem like the easiest way to manage people. It is, however, damaging to your company. People are creative and will game your system and work on things that do not benefit your organization. What is even worse than the loss of productivity, however, is that you may end up sabotaging yourself and punishing or getting rid of your best people. Unless you truly are a sweatshop, where you have a quota of how many gizmos need to be produced every X number of hours, it is usually a mistake to measure humans with numbers.

One of the reasons metrics are appreciated by management is they are a non-emotional way to recognize and deal with problems. On a high level, that is a good strategy. On lower levels, such as individual performance, however, metrics are useless unless taken in context.

We can start by looking at typical scenarios and how they miss the mark in Information Technology.

It may appear logical that the more code a developer submits, the better the developer. In the old days, that might equate to lines of code. Now it might be number of commits. A commit is an update of new/changed code to a common repository. You may also feel the better the developer, the less errors will manifest in their code. So you may measure both number of commits and errors attributable to each individual.

Now let's look at those assumptions in context. It is likely your best development staff will be assigned to the most difficult areas of your applications. If that is true for you, no matter how good your developer might be, the nature of the design, architecture, and coding will take more time and be more error-prone. If you punish your development staff on the basis of these metrics, you will get several bad behaviors or results.

First, you will get plenty of commits that won't necessarily do what you would hope they would do in as efficient a manner as possible. Second, any error found by a third party will be hotly and violently protested and refuted, losing your company time, money, and effort better spent on other things. If you have a QA department, they will hate these metrics as much, if not more, than your developers. Your development staff will actively avoid taking on or accepting work for anything even remotely complex. Do you want to encourage those behaviors? Do you want your best people to be fired, leave, or push back on taking work in keeping with their level of expertise? I doubt it.

Let's look at some typical measurements that negatively impact QA – which is normally your testing staff. You might reason that the more tests an analyst designs, the better they are. You might assume the more bugs (errors) a person finds, the better they are. You may choose to measure both of those things.

Now consider those numbers in context. The number of tests an analyst designs depends solely on what needs to be tested. Some things will require two tests. Some will require five hundred tests. It has absolutely nothing to do with the skill of the analyst. If what you require, however, is huge numbers of worthless and meaningless tests, by all means use that kind of metric to try to measure individual performance. People are not stupid – they will game your system.

The number of bugs found by an individual is highly dependent on the application(s) under test. If they contain a lot of bugs (errors), the tester will find a lot of bugs. If the development team has done an awesome job, they might find only a few. If, however, you measure individuals by the number of bugs they find, you're going to get a lot of spurious, non-bugs written that will take your development teams time and effort to sort through, and everyone on the QA team is going to want to work on the sloppiest, most error-prone parts of your system, leaving work on other more robust (and possibly more valuable) parts of your system in jeopardy.

Again, using these particular types of metrics, you might ding a QA person for things that have absolutely nothing to do with their skill, experience, dedication, or anything of value to your company.

It is totally understandable why management prefers the non-emotional and easy measurement associated with numbers, but every number associated with human work performance needs to be taken in context. If you look at those measurements logically, in order for them to have any validity everything would have to be equal. The skill and experience level of the individuals. The complexity and size of the work. Those things are never equal. Each person is at a different skill level (and likely being paid accordingly) and working on something different.

Unlike many of my IT brethren, I love metrics. Metrics, applied correctly, can tell you a story you can't get any other way. At the same time, they have to be captured and applied intelligently to provide maximum benefit to the company.

Can they ever be used to determine individual performance levels? A better use would be to use them as an opportunity to mentor. You can't compare a newbie to a senior and expect similar results from numbers. Numbers can point out an issue or a potential problem; it's up to you to put things in context and figure out the "why".

Your metrics become more and more useful and interesting the further away you pull from monitoring individuals. Oddly enough, pulling the right numbers for the right reasons can eventually assist with individual performance, but not when you start for those purposes. You end up trying to look at root causes without understanding the real problems and impacts first. Have you ever read The Hitchhiker's Guide to the Galaxy by Douglas Adams? They wanted the answer to life, the universe, and everything. The answer was, after 7.5 million years of consideration by the supercomputer, "42". When they were baffled by the answer, the supercomputer told them they had to understand the question first. Metrics are like that.

When you first start out collecting metrics, your purpose should be to establish a baseline. Once you have that, you know where you are right now, at this point in time and if you need to do anything about what the numbers tell you.

The numbers you collect need to be gathered the same way each time in order to be useful (and honest), tell a story, and they need not be sophisticated when you start out. You need to know what numbers you'd like to have and what would be actionable and useful. Otherwise, whatever you gather will be largely useless and you'll be trying to find a use for numbers you really didn't want or need.

I'm going to give examples from the software quality space. For purposes of this example, let's assume you are an Agile shop, and you produce code that moves to Production every 2 weeks.

How many errors manifest in Production once the code is made available to your customers? How serious are the errors?

This is an incredibly simplistic number and counting the issues and the severity level of those issues is likely already documented by your customer service folks, so they are easy to gather.

Take a look at these numbers. Are they pretty much the same every two weeks? If not, you have to do a deeper dive. This is where taking the numbers in context comes in. If the numbers varied widely between one deployment and another, you have questions to ask. Why was one number higher than another? Was the technology new? Were there new people on the team? Severity levels are important because metrics detractors will point out it doesn't matter how many errors slip through to Production if one of them blows your business (and thus your revenue stream) out of the water for a period of time. They're right about that.

Say your company logged 20 errors in Production and of those, none were high severity levels. Now say your company logged 5 errors in Production and of those, one was so bad your customers couldn't use your applications for a day and a half. Which one of these was the worst scenario? The answer is it was the one that impacted your revenue.

However, every error found in Production is a failure on the part of the development team. Every error, big and small. In the Quality Assurance world, you want people who are so anal about their jobs they take every error that slipped by them and made it into Production as a personal affront. They should be all over these. Did they have a test for this? If not, they should add it to the test base. It might have slipped by them once, but it won't slip by them twice. If they did have a test, did they run it? If not, why not? Were they told not to? If so, what did they (and those giving those orders) learn from the experience? Did a tester skip it on their own recognizance? If so, that was an error in judgement and it gives you, as a manager, an opportunity to mentor them. If it was a serious error, you might be tempted to mentor them upside the head. Resist the temptation.

If an error makes it into Production, consider how it arrived there. A developer coded that error. They should have tested their own stuff before passing it on and they missed it there too. They handed it off to a third-party tester (your QA staff). Who also missed it. It is also possible the error was found, but the team decided not to fix it. If it impacted your customers and they contacted your Customer Support about it, that was an error in judgement.

Overall, there are at least 3 points of failure in every issue that makes it to Production. To add to the situation, your production environment might have differences from your testing environment that literally makes it impossible for your staff to locate and fix certain types of issues. All of these make production errors an opportunity in disguise. Use them to get funding, time, and attention paid to improving your development and testing environments. Use them to improve development and testing processes. Use them as a good way to guide and mentor your staff – both developers and QA.

When considering software quality, it isn't a bad idea to set quality goals for your company. It might be that the error rate (more on that later) has to be below 10%, and of those, zero are Severity Level 1 problems. Errors are usually categorized by severity. A Level 1 problem is an error that impacts your customer base and renders some aspect of your applications unusable, unstable, or dangerous. If you are new to software quality assurance, a top-notch quality program with great people is going to get you an error rate under 3%, with severe errors well below 1%. Why does this matter? Because if you stand at about 20%, your staff are going to be spending much of their time frantically repairing, retesting, and redesigning old stuff, rather than working on new ways to bring in revenue. If you stand close to 40%, your apps are going to be unreliable and potentially unavailable. You'll move code to Production and everyone in the company will be holding their breath to see what happens.

The advantage of setting quality standards in advance is that it cuts down on the time and effort spent by your teams trying to decide what does or does not need to be fixed in order to move code to Production. Anything that saves time equates to saving money and is a win.

After you've gathered metrics every two weeks for 2 or 3 months, you should be able to see some sort of pattern or indication of what is "normal" for you. This is your baseline. You can now use those metrics for a variety of reasons. You can set goals to lower those numbers. You may decide those numbers are fine and set them as your standard and not a concern unless you see a trend where the error rate or severities are increasing.

Once you have a basic like this under your belt, it's likely you'll want to start capturing other statistics that are even more valuable to you. That is completely normal and a very good way to start your metrics journey. Start with the lowest common denominator you have and when you have a handle on that, add some levels of sophistication.

Again, in the QA space, what I, personally, find valuable are error rates. This requires counting the number of tests run and the errors found through that testing. The equation is (number of errors found)/(number of tests run) X 100. This is the percentage of error. For example, if you run 120 tests and find 5 errors, your error rate is 5/120 * 100, or 4.1%.

You can extrapolate further by noting severity levels. If, for example, there were no Level 1 errors found, the percentage of Severity Level 1 bugs was 0%. If you found 2, the percentage was 40%. In other words, 40% of all errors found were Severity Level 1 bugs. The formula is (number of a particular Severity Level bugs found)/(number of overall errors found) * 100. You need to take even these metrics in context. The lower the number of errors found, the higher the percentages of whatever severity levels were found. If I find only 2 errors, for example, and one is a Severity Level 1, 50% of all errors found are Severity Level 1 bugs. This can be misleading.

I can hear exploratory testers out there screaming from here. Exploratory testers don't necessarily document tests prior to receiving the software they need to explore and believe strongly that human beings, with their superior observational skills, actually notice and observe potentially hundreds of things as they are performing a "test". They are not wrong.

There are problems with that viewpoint, however, depending on how they choose to perform their testing and the environment and products in which they typically test. While it certainly causes problems for collecting meaningful metrics (you can't count what you don't have), there are other problems with operating this way that are much worse. I do want to point out that inevitably testers who do not (or do not like to) document tests say they are "exploratory" testers. It is not necessarily a true characteristic of exploratory testing; it is merely being used as an excuse to skip activities that aren't much fun.

And testing is fun. There's a reason I still love to test after all these years and it isn't because testing is boring. Where the rubber meets the road, however, is the discipline and professionalism of your staff. Documenting tests is NOT fun. If you're a professional and not just a hack, however, you are going to document your tests in some fashion because you understand the necessity and that activity is going to take place as early in the development process as possible.

It is entirely reasonable to expect a professional Quality or Test Analyst to be able to look at a page design or any kind of spec

[187]

and to be able to list the tests they plan to run immediately. Will that list, structured outline, or "test plan" be cast in stone? No. It will change as changes come up. During the testing itself, when all of those lovely observational skills get put to use, they should be expanded. The benefit of starting with your intelligent list, however, is that you, as a human being that might have limited space in their head or be exhausted after a hard day, will not inadvertently skip something you meant to look at based on the specifications. Further, it protects the team and the company. If you get sick, injured, or leave, any member of the team or your QA compadres should be able to take your list and figure out what you, as the testing expert in that area, planned to examine.

It can be borderline catastrophic when someone who is an expert gets swept away by something just when you need them most. A test outline, "test plan", or intelligent list protects the company from single points of failure that could have been avoided by this process. Lack of this basic documentation also wreaks havoc with development teams (who won't know what tests they need to pass), and any QA associates that might have to pick up the slack when someone is sick or absent for any reason. It can put test automation efforts even further behind than normal.

Documentation, at its very basic level, can also save your butt when it comes to any kind of audit. Even if you are not subjected to periodic audits right now, as you grow that might be in your future. From an individual QA perspective, documentation is not optional; it is a requirement of the job and it needs to be respected and produced for the protection of the individual, the team, and the company.

Personally, I do not think it reasonable to require stepped-out test cases during the development process itself. They take too long to produce. They can be expanded later and/or handed off to another team for expansion. By the way, if you want stepped-out test cases (which are strongly preferred by test automation staff), this is a good activity (expansion of test cases) for newbies, as it requires them to learn the system, or for making use of off-shore teams (who often switch out team members, are in a different time zone, and may not have the skill levels of your regular staff).

Better yet, let's all keep an eye on AI and see if it can do this time-consuming and strongly disliked task for us sometime in the future. It's not quite there yet.

Are you convinced you need some basic documentation? Good. Let's move on.

If you gather statistics regarding the number of tests run and the number of errors found with their associated severity levels, you now have a goldmine of information and metrics that can make a difference. Metrics are actually useless in and of themselves until they are placed in some kind of context, but once you understand the context they are an outstanding way to allow you to ask the right questions and point out problem areas.

Let's take a real-life example to illustrate this. I worked with a company that had serious problems with errors in Production. I've worked with many such companies – generally companies don't hire at my level unless there are quality problems to solve. Their staff were unfocused, overworked, and burnt out, nothing was done on time, and their errors in Production were affecting both their bottom line and their future plans. Both development and QA staff were working tons of overtime and there were strong feelings by development that QA took too long. This is a common complaint. The other areas of the company weren't too impressed with development efforts, period, since the lack of quality impacted their lives as well.

Using the statistics above, I found their error rate in Production was approaching 40%. This is bad news – it's difficult to work on new stuff when you're busy repairing the old stuff.

So I drilled down further. The error rate in the test environment (when development turns over their code to be tested prior to moving it to Production) was over 300%. That means for every test run (remember that tests are run at the end of the development process), 3 or 4 errors were uncovered.

Well.

That number is about 260% higher than it should be. Everyone was dumbfounded by those numbers, but it explained why everything was going sideways. What did we do? We collaborated. We communicated these numbers to the team and explained our goal was going to be 40% in the test environment and under 10% in Production. We did not get arguments and pushback from executive management strategizing our solutions. That's another advantage to using non-emotional numbers. You can (and likely do) argue with a human being. You can't argue with numbers. They are what they are. We did not get arguments and pushback from the team(s) involved. They could clearly see improvements needed to be made.

What we gained from this process? More staff. Changed development processes, which cut back on the number of tickets (chunks of work) being tackled by each developer. Some switching around and hiring of senior people who could help out on the technical side. A focus and insistence that each chunk of work be tested (unit testing) prior moving everything to QA. Better test environments for both development and QA staff.

What did that company achieve? It wasn't instantaneous. But their error rate is around 1% in Production now. Severe errors are rare. Development gets more respect. As an aside, each time the team(s) improved, their improvements were highlighted and applauded. Although we could have driven down even deeper and attributed X number of errors to Y person, we didn't. Punishment is not especially motivating. Goals and acknowledgement of achievements work better.

What those numbers did is give us a jumping-off point for asking the right questions. Continuing to monitor those numbers told us when we were on the right track and improving, and when we fumbled and went backwards. It gave us something concrete to take to upper management to get funding for improvements and people. It gave our teams goals they could understand and support.

Metrics can do that for you too. What's more, they can, when gathered using some regular cadence and in the same way every time, show you trends that are valuable to your strategizing. They can tell you when changes or improvements have made a difference (whether good or bad). They can help you decide where to focus your efforts and your money. They can help you establish goals everyone can understand and support.

[191]

Detractors are going to tell you that the very nature of metrics make them rhinestone processes; you can make numbers say whatever you want them to say. With some expert practitioners, that's close to true. There are people out there that can make metrics jump through hoops in order to get support and/or funding for something they want. Usually those people are pretty high up in the corporate food chain.

My thought is "So what?". You have no control over some treacherous troll in some other area. You, personally, can do better and use metrics to benefit your staff and your company. You really can't tell where you want and need to be unless you have some idea (other than a "gut feeling") of where you are right now. Start with obvious, easy numbers. If you're in software development, that might be the number of tickets pulled into a sprint and actually completed. After you look at those numbers for a few months, you'll see a pattern and have a baseline of what "normal" looks like. Then you can take it from there. Is "normal" OK, or do you need to improve it? How? Using metrics to help you strategize is one of the reasons they are appreciated and used by management staff.

If you're in management, aspire to management, or need to get something from your management, you need to swallow your doubts and dislike and learn to establish and work with metrics. They only become rhinestone numbers if you're trying to tell a story that isn't true or use them to measure something with no context. Used as just another tool in your toolbox, they can help you and your company keep a handle on what's going on, strategize, allocate people and funds, set goals everyone can support and understand, and be more successful.

CHAPTER FOURTEEN
QUALITY OWNERSHIP

This chapter is a bonus. It was presented in 2023 at the Pacific Northwest Software Quality Conference (PNSQC) and won a Top Paper Award. I've since updated it to include more information and suggestions for AI. No doubt it seems impossible that so much has changed in two years, but that is nothing new for technology.

I'm including these ideas in this book because it offers solutions to several common software quality problems – everybody talks about them, but as is common with rhinestone processes, no one ever does anything about them.

If we're ever going to get ourselves out of rhinestone rhetoric that sound good but don't do anything for us (or anyone else), we need to move forward. 40 years of the same-old, same-old isn't cutting it.

In order to understand the ideas behind Quality Ownership, a basic understanding of the history of the field might be in order. This is a brief overview of the quality world during the past 40 years or so, told from the perspective of a practitioner.

40+ years ago, everyone was using Waterfall methodology. Many modern perceptions of what waterfall was like are, well, just wrong. If you've never lived it, you need to stop talking about it like an expert. You're going on hearsay and anyone who has actually worked on Waterfall projects is going to look at you as if you've grown two heads.

Final implementation to Production might have been "Big Bang", but large projects were broken down into phases, often a month long. You might compare it to a 4-week sprint.

At that time, in my world, QA staff were developers who reviewed code, made structural updates, ran tests, and fixed bugs. This was done at the end of the phase and again at the end of the entire project.

That final review/test phase could bog down for months. QA was considered a bottleneck. The mission of the QA team was considered a failure if errors were detected in Production.

The QA world changed in the early 80s, when a book that had been around since 1973 started getting some traction – Glenford Meyers' "The Art of Software Testing". It talked, for the first time, about the psychology of finding errors; pointing out realities that still ring true today.

For example, it talked about WHY it's difficult to find errors in your own code. Consider, for example, that I write a report for the CEO. I want it to be well-received. I may read over and check my work ten times. I will find and correct some errors. If I then ask a colleague to read it just as a final check, they will find multiple things I missed – some of them glaringly obvious. Why did I miss them? The reason is psychological.

We don't really want to find errors in our own work. In essence, developers test to prove their code works. QA tests to prove it doesn't. Just that difference in perspective drives out error.

For me, the biggest takeaway was unless you are writing code for other developers, your clients do not care what language your code is in. They don't care about your environment, your methodologies, or your expenses. YOU care about those things. They just want something does what they want it to do and doesn't break. Now there's a simple concept of "quality", but in terms of software applications, it pretty much covers it.

So the field morphed and QA as a whole started testing systems from a user perspective. While this certainly was faster, responsibility for structure, bug fixes, and everything but the actual testing devolved to the development staff.

Tests were defined by QA as soon as the analysis document was available, which took time, and as code that was often untested and incomplete was passed over to QA staff for testing at the end of a phase or project, the QA process overall was still time-consuming. QA was still considered a bottleneck. Errors in Production were still considered a failure on the part of the QA team.

Enter Agile. Agile methodologies came out in 2001. What is really interesting about Agile is that it's 24 years old and everybody is Agile. Well, everyone SAYS they're Agile. Many of the original ideas behind success for the original agile teams have simply been lost as our field continues to evolve. Do you have a closely-knit, co-located team in the same time zone? Have someone that protects the team and does not allow them to be interrupted? Do they pitch in and help each other out? So they're all full-stack developers? Does the entire team do the testing?

If not, you may have something and it might work well for you, but you're not really close to the original concepts behind Agile.

And what about QA? How did their world change? Here's the bad news. It didn't. It's just in 2-week chunks instead of 4-week chunks.

With the advent of exploratory testing, which most companies still think means "no documentation" or preparation, our terminology changed, but the methods we use to uncover and report bugs really didn't change.

If there is a QA person on the team, they are likely responsible for testing, regardless of when or how things are delivered to them.

Chances are also good QA will be responsible for test automation, and if something fails in Production, it is equally likely the QA team is getting the call, "Why did you miss this?".

In a truly Agile shop, this would be viewed as an agile process error; everyone is responsible for testing and the quality of the project. But in reality, QA is still considered a bottleneck when the team fails to make a deadline and a failure if errors manifest in Production.

There have been and are many flavors of project methodology, but the mission and methodologies of a quality professional have not changed significantly since the early 1980s. RUP (Rational Unified Process), MBO (Management by Objective), Six Sigma, Lean, Scrum, Kanban, TDD (Test-Driven Development, BDD (Behavior-Driven Development) and all derivatives thereof – it doesn't matter; a QA professional needs to be able to support any type of project methodology and provide value to their teams and their company. Our job is to adapt.

From process-heavy methodologies like RUP, to admin-heavy processes like MBO or Six Sigma, to the ceremonies and time-boxed Scrum methodologies and the CI/CD (Continuous Integration/Continuous Deployment) -friendly looser weave of Kanban, all of these methodologies can and have been successful. Methodology of any type can and has been unsuccessful as well.

Overall, success of any given team is more about the team itself. A talented, invested, dedicated, supportive and supported team is going to make things work in spite of obstacles. So why work on methodologies? Because we can and do hamstring our own teams, burn them out, and demotivate/demoralize the very people we depend upon.

Let's consider the nature of testing itself. While a test can be developed before anything actually exists to run it against, the testing itself cannot be performed and give you any meaningful information until there is something to test. "Test early, fail often" sounds great and all testing is valuable, but the errors find as you shift left (testing whatever you can as early as you can) are different than those you find later in your development cycle prior to production.

My code may be tested and completely performant, yet the nanosecond I put it in place and it interacts with other code in a more production-like environment, it can fail. It can cause other code far down the line from my own to fail. Overall, it is the later phases of testing that are most indicative of performance in production.

It is time, after more than 40 years, to recognize a primary truth about testing. It takes place at the end of something. It takes place once something done. The most meaningful testing takes place at the end of a development effort when all of the pieces are ready to interact with each other in an environment as close to production reality as possible.

The reason testing extends a deadline is that its very nature demands end-of-cycle execution. The purpose of testing is to find errors, all of which have to be assessed and potentially fixed by the team. This takes time and effort by human beings. It is true whether the tests are automated or manual as results must be analyzed, assessed, and potentially fixed by a human.

This testing reality is true regardless of methodology. So, the first challenge in terms of change for the better is finding ways to make an activity that can't take place until something is done faster without losing any of the benefits of that activity.

Diving even deeper, what is "testing"? Testing is comparing a given result against an expected or desired result and noting anomalies so they can be repaired – either prior to Production or later. Or the decision may be made that the anomaly is acceptable as it stands. How is this decided? On other words, who decides what is good?

From a practical perspective, eventually it will be your customers that decide what is good. But using our simplistic definition above, your product does what is expected and doesn't break; the expectations against which to compare the results of your testing should be documented somewhere.

In Waterfall days, these expectations of what needed to be produced, what it should look like, and how it should operate were in a bulky analysis document. This document took a very long time to prepare, but it did have some advantages. First, everyone on the team had a complete picture of what was desired before coding or testing started. Second, it was so cumbersome to create, changes were discouraged and normally had to go through some thorough vetting and a change process before it was accepted.

This was both a blessing and a curse. The blessing was that spurious and unnecessary changes and bells and whistles were not added as often, which kept disruptions to the forward momentum of the team(s) to a minimum.

It was a curse in that the teams were far less responsive to changes in the marketplace. Technical issues or discoveries later in the process caused longer and more stressful updates to what had already been done, with the added burden of ensuring everyone on every team was kept in the loop about those changes.

Now most analysis work is kept in a variety of mediums. There might be one site for designs and the work is likely broken down into pieces ("tickets") in a project management tool such as Jira. Jira tickets and similar contemporary project management tools are also both a blessing and a curse.

The blessing is that a ticket represents one small element of work and is fast and easy to produce. Changes are easier to absorb. While a change might have to be vetted, it might be by a single team and only to determine if it can be handled during the current sprint or needs to be placed in a later sprint.

There are a number of curses. Reading a single ticket does not give a very good picture of the overall effort, resulting in more rework and change later down the line. Changes may not be vetted as extensively for impact to the product and team or even if it is desirable from a user perspective because the "approving body" is much smaller and far more focused on their own specialized tasks.

It is also unusual for a ticket to be detailed enough to use effectively as a basis for testing. Tickets are "supposed" to contain acceptance criteria. Often, they do not. They may contain generic statements that are a re-iteration of the title or intent of that piece of work, but acceptance criteria go far beyond that in reality.

For example, if you are creating a log-in screen, your acceptance criteria might simply state the user is able to log in to your application. A testing professional is going to have quite a few tests associated with such a ticket. What happens if the log-in is wrong? If an incorrect log-in allows access to your application, that anomaly, or bug, will be written up and the team will have to address it. In other words, the acceptance criteria is actually every test which if failed will result in a bug which must be addressed by the team.

So who is responsible for determining the test set for the ticket? The Product Owner, Product Manager, or BA might be responsible for creating initial acceptance criteria. But it is normally the quality professional who is responsible for taking any design documentation and the ticket to determine what needs to be tested.

If the QA personnel on your teams do not begin this work until a ticket is pulled into a sprint to be worked, or they do not document what they intend to examine, your agile team is already behind.

The critical analysis piece of the equation will be started at the same time as development. There is no way they will be able to use those tests for their own unit/shift-left testing, and getting automation done prior to the end of the sprint will be problematic.

Ultimately, the test analysis is the most critical part of the testing process. It doesn't matter what kind of analysis documentation you have (or don't have). It doesn't matter what kind of project methodology you use. The key to ensuring a valuable testing effort is running the right test, at the right time, in the right environment and providing the results to the team.

Shocking as it may seem, it makes no difference whether that test is manual or automated. It doesn't matter who (or what) actually executes the test. The key is running the right test, at the right time, in the right environment.

This means the most important part of the entire testing process is test analysis. Determining what needs to be examined and comparing actual results to expected/desired results. Good testing is always a comparison.

There is a big difference between executing code and verifying code. When there are no expected results, it is not possible to ascertain whether a result is desirable or undesirable except through the experience and expectations of the tester themselves.

This will work well if your tester is highly experienced, has extensive product knowledge from your users' perspective, and is 100% available. They will make good guesses most of the time, although a chunk of time is going to be spent consulting with others to get their perspective on any perceived anomalies. But it represents a risk to your company and is ultimately not sustainable.

Most companies would not want the future of their organizations dependent on guesses, yet many operate this way with varying degrees of success or lack thereof. It is also important to the team that this test analysis is done as early in the process as possible. This allows development staff to understand and fold testing expectations into their work and the entire team to agree on what 'done" looks like.

Thus far, we've determined that testing takes place when something is done and that the key to successful testing is good test analysis done early. Let's talk about people next.

There are many myths around the testing practice. Everyone who has ever tested anything, including a battery, believes they know everything about testing, that it's easy, and that the entire field is somehow a lesser activity that development. In actuality, god testing requires training, experience, and discipline. Testing is part of the development process. The specialized nature of the field should be evident; there is a reason QA professionals find so many issues after development is complete. Why aren't developers catching these errors? Where is the disconnect? Is that perception that's been around for over 40 years that "developers are terrible testers" really true?

The short answer is no. But let's talk about the disconnect. Say my company is an on-line marketplace. We add vendors to our system, which enables them to make their goods available to customers who can purchase said items on-line. A QA professional is going to know how to add a vendor, add goods, access those goods as a customer, purchase goods using a variety of payment methods, and verify the vendor and company got paid. They may also check any analysis/reports provided to the vendor or receipts to the customer. They are going to know these things and more because they have to in order to do their jobs.

A developer, on the other hand, is likely to specialize in either front-end or back-end work and be deeply, technically involved in one small part of that chain. Generally, people learn and do what is required for them to be effective and no more; most people don't have the time (especially in the IT field) even if they have the interest. There are developers out there doing good work and contributing to their companies that have never even logged on to their own systems as an end user. Does that mean they are incapable of learning? No. But it does mean they lack the breadth of understanding of their own systems from a user perspective necessary to do good test analysis.

Think about the best developers you've ever known. You know the ones – you call them at 3 AM when something goes awry, whether it's actually in their scope of responsibility or not. It is highly likely the development staff most valuable you are those that know your system from BOTH a technical and a user perspective. A customer support person can explain what user did and they understand instantly and ask questions customer support understands, then can pinpoint the area of code most likely at fault and engage their cohorts in finding a solution. This level of employee is going to be better at test analysis – unfortunately they are rare and unlikely to be used for such a purpose.

What this means is the disconnect is due to the learning requirements and time available to gain the expertise necessary to be an effective test analyst and not capability. It is again the test analysis part of the equation here, not the actual testing. The actual testing activity does not require any particular skill. It can, after all be run by a machine. Anyone with time available and understanding of basic skills can run a defined test.

Understanding what test needs to be run? That's a different story. Test analysis is a highly specialized skill.

What about other members of the team or the company? What about Product Owners? Product Owners certainly understand the system from a user perspective. What they lack is training and time. Does your Product Owner/Manager have the right skills to perform test analysis? Take a look at their tickets. What does their acceptance criteria look like? Do they provide any? Do they include how the system handles invalid or unexpected input? Do they mention downstream applications that need to be verified as part of the change? If not, even your lowest-level Quality Analyst is going to be more effective at test analysis.

What about an SDET (Software Development Engineer in Test)? This is an interesting question because there are so many assumptions and myths surrounding test automation as a whole. There's an assumption that test automation is a recent development (it isn't). Every company, everywhere, always want their test banks(s) automated.

And why not? Something that takes manual testers weeks to execute can take less than a day, even with the human interaction portion of reviewing test results and failures. But in order to automate a test bank, you have to have a test bank. Again, who is going to design your tests? An SDET? They just want to automate tests. They are developers, not test analysts.

They interact with, think like, and work like development staff. They tend to miss the same bugs development staff do. That makes sense, because their understanding of your systems from a user perspective is likely on the same level as your development staff.

A common myth in the IT field is that an SDET is just a Quality Analyst, except they can also automate tests and are therefore a better investment long-term. Unfortunately, this is just not true. Even if it was, an SDET with years of experience is going to be expensive and generally they will not enjoy and do not want to be involved with any significant tasks that do not involve automation specifically.

Where are we now? Thus far, we've determined that testing takes place when something is done, that the key to successful testing is good test analysis done early, and that Quality or Test analysts are best positioned to perform testing analysis. The last reality to tackle is process.

It is to be hoped that any company interested in innovation and with some claims to operating in an agile environment periodically reviews its own processes, identified those that are holding it back, and either provides training or tosses them accordingly.

That is, unfortunately, a hope in vain. People resist change. There is some comfort in familiarity, and fear of losing respect bv not being expert (or being unable to be competent) in something that is new.

In addition, there is a warm, fuzzy feeling when you are part of a pack or you do things the same way those other 2000 reputable, successful companies do. That is not, however, necessarily in your best interest. Instead of being a follower, you could be a leader. And leaders do what is in the best interests of their own company. Not someone else's. We need to look at our processes and be tenacious about cutting what doesn't work.

Let's look at a few agile processes for a moment. If you do not have a tight, co-located team in the same time zone are you agile? If the answer is no and you're successful, who cares? You can call your process anything you like and laugh all the way to the bank. If the answer is no and you're struggling with processes that do not work for you, it's time to take a second look.

Are all your team members in the stand-ups or are some in a time zone that does not permit that? How do you handle that problem? Is it really handled, or are some team members unavailable or not in the loop? Are your retrospectives more than a 4-syllable word? What valuable changes have you made in the past 6 months due to retrospectives? The idea of a retrospective is great. Lots of ideas are great. But unless you appreciably benefit from an agile ceremony, let it go.

What about RCAs (Root Cause Analyses)? Do you do them? Are they about improving processes and tools or providing training or capability growth, or do they simply blame people? Do your people embrace them, or do they dread them? If the answer is b., what are going to do about it? And if you've been doing these same ineffective, damaging things for years, then just stop. Achieve your goals in some other way.

With this in mind, let's look at a typical scrum-like project sprint. A change or new feature is identified. Any new/changed screens and workflow are worked on and placed in (some) tool. The workflow and changes are broken down and estimated by a team(s). Including QA time in estimates may or may not be included. The Product Owner pulls tickets into the next sprint based on team capacity. Team capacity may be based on historical data or may not.

The development staff starts working. The QA staff begin test analysis. Once an individual ticket is complete, it may be tested by either the developer or by QA in a non-prod-like environment. Bugs are written and the fix/retest cycle continues.

Once all the tickets are complete or a working feature is available, the QA staff retests the entire feature in a prod-like environment. All needed tickets may not be available at the same time, resulting in some delays. In addition, several design changes might take place coming directly from the executive team. Errors in the estimation process (perhaps using story pointing) become evident, but the team is committed now and moves forward. The fix/retest cycle at this point in the process strongly impacts team delivery dates.

Once the delivery date has been reached, the team schedules its next sprint, with QA staff taking time out of their testing (and developers of their fixes/resubmissions) and development begins for the next sprint, although the last one is not complete and in production. If enough time passes during the final fix/retest cycle, two sprints may be combined. Team capacity may or may not be updated accordingly. Once the sprint is in Production, tests are passed off to SDET personnel for regression automation.

Does any of this resonate with you? This process is not effective. We need to address when test analysis takes place, when a ticket is pulled in as "ready to work", how tickets are estimated, when tickets are automated, who is responsible for the physical activity of testing, and how team capacity is determined. Note there was no mention of addressing design changes coming from the executive team. If this is part of your reality, a better perspective is to ensure your final processes support these as a 'business as usual" part of your landscape. Admonishing an executive team is not an effective strategy.

Before we continue with a potential solution, the above examples are IT-related, but there needs to be some discussion regarding quality as a whole. While the most simplistic definition of quality is that it does what it is supposed to do and does not break, there is actually more to the concept of "quality" than those two items.

It is not a matter of who has the prettiest website either, although doing your homework and understanding what your customers want is important or no one will buy what you're selling.

When someone thinks about quality and whether your company has it and how they "grade" you on a variety of sites is dependent on absolutely everything that touches them. You can have a killer app that has zero errors, but if your customer service rep treats someone poorly or cannot solve a problem, you've lost a customer. If you have any competition whatsoever, you have to pay attention to and "test" everything that touches your customer. Your website. Your social media posts. Definitely your customer service practices. Emails. Texts. Mailings. Everything that touches your customer influences their perception of the quality of your organization.

This is a basic reality. Even if your customer base is locked in and has no choice but to use your apps, that could change in the future – best to treat them like the gold they are right now.

In summary, we now know that testing takes place when something is done, that the key to successful testing is good test analysis done early by the Quality/Test Analyst, and that our process is broken and needs work. We would also like our solution to be useful to any part of our company that touches our customers.

Let's talk about potential solutions to these problems.

There was a conference for quality professionals back in the 1980s called "Quality is Free". Quality is not and never has been free. Quality is expensive both in time and resources. It is, and always has been, difficult to justify and obtain sufficient quality professionals to handle the workload. The only way to support growing applications and company size has typically been to expand the QA presence accordingly, which has always been problematic.

What if we took a page from the book of other specialists on agile teams and made it possible for a single Quality/Test Analyst to support more than one team? DevOps personnel, for example, might support multiple teams, as can Product Owners, UI Analysts, and others. What would that look like for quality professionals?

First, the single most critical contribution a Quality Analyst makes to a product team is establishing the test base – defining what tests need to be run. That is a task that others on the team cannot perform as successfully. At the time of this writing, AI is not yet capable of designing good or comprehensive functional tests. It can, however, be more useful for formulating unit tests.

If we agree Quality Analysts should establish the test base with or without the assistance of AI tools, when and how should that happen? If we want to solve the problems of not having tests available to be incorporated into shift-left testing efforts, having tests automated by the end of the sprint, and development beginning their work potentially weeks before QA begins theirs, it makes sense to say the QA Analyst will establish a test base prior to a given ticket being pulled into a sprint as "ready to work".

A separate book could be written just to deal with the "hows", and it's highly dependent on how a company operates now, but this would mean involving the QA Analyst during the design phase to define the functional test base, employing AI to determine unit test sets, and ensuring all tests are documented, stored, and attached the ticket prior to being pulled into a sprint.

I use the word 'sprint", but if you work with another methodology, it would be prior to whatever block of time you divide into work chunks.

Once the development staff pick up a ticket to work, there are two possible scenarios. The original idea was to hand the ticket over to an SDET for test automation at the same time the developer pulled it in to begin work, but I had an interesting conversation with a development director I was working with at the time that turned things around for me.

He had a team with no QA resources on it, producing some work for internal customers. To be frank, the QA team, in particular the QA Analysts, were a bit chary of him; they felt he didn't recognize their value and were afraid he wanted to get rid of QA altogether.

In actuality, he had a high degree of respect for the Quality Analysts and credited them with a great deal of value in designing tests and finding critical bugs. What he felt was overkill were the SDET (test automation) staff. That is a relatively astounding, ground-breaking, non-typical point of view. What he said was that he had a whole team of developers. With an established automation framework, any one of them could automate a test. What he did not have, and wanted, was all of the experience and understanding of the systems from a user perspective helping his teams determine what needed to be tested.

In other words, he wanted test analysis. From there, he felt his team(s) could take over.

Regardless of whether this task is assumed by an SDET or a developer, automated tests can be developed before code is available with most contemporary automation frameworks. That is one of the foundations of TDD (Test-Driven Development). Once elements/objects and actions/functions are defined, these are easy to build in a basic form, although more tweaking may be necessary as the code develops and the tests are actually run.

As part of the development efforts, time to automate the tests would necessitate considering them as part of the estimation process (a process often skipped now), improving the track record of the team for on-time deliveries.

In addition, the availability and running of automated tests, incorporating them into CI/CD processes, would benefit the entire team in terms of time. It would encourage teams to automate, and automate early, which is very much in keeping with the direction in which most companies want to move.

One of the biggest benefits, to my mind, it that it provides a significantly better ROI (Return on Investment) for test automation. Test automation is better at unit testing than individuals. It is not better at functional testing. Yet due to current process scenarios, where they are last in line to get the documentation they need to do the work, their roles are relegated to creating automated tests after code is in Production, This means most companies lose out on the primary benefits of test automation.

The most critical and controversial part of Quality Ownership is the question of who actually performs the testing. One of the primary reasons "testing takes too long" is you typically have a single quality expert trying to test new stuff, changes, and perform regression testing (new stuff doesn't negatively impact existing stuff) in a short amount of time.

I'm going to do you a huge favor. Your testing team is about to triple or quadruple in size without you spending a penny.

On an agile team, every single member of the team is responsible for testing. The key point to success for the entire Quality Ownership process is "everyone tests until the code moves to Production". This is a non-negotiable. Why? Because breaking that "QA will test it" mentality will not go away and the teams will not grow and move forward until that crutch of handing all the work over to one person is no longer available.

I saw a podcast that talked about a situation where a QA leader had a team that was being blamed for a production failure and she had no people to spare to help them out any further. The development leader was furious. But reality is reality and his team was forced to carry on with no testing support. Guess what? They became better. No amount of QA professionals on the team had improved the quality of their output, but once they were responsible and assumed responsibility for the testing, they were forced to improve.

The idea of expecting the team to test their code also forces other staff to learn how to use their own product from a user perspective, which grows highly expert, valuable personnel. The available test pool grows from one person who is often trying to 'save' the team by testing overtime at the last minute, to the entire team, all working together to get the job done. That is the epitome of agile methodology.

Again, with testing genuinely part of the development process, estimation techniques would necessarily have to improve in order to allow for the time to execute the test/retest cycle. And those valuable, expensive QA resources can support more than one team. If their primary job is to identify the test set, once that is complete, they can move over and help do the same thing for another team.

How does this relate to non-IT teams like Customer Support, Marketing, and others? Anything that touches a customer should be "tested" prior to presentation to that customer. This would be a simplified process; a Quality Leader would identify what needs to be verified and ensure those steps are done and acceptable prior to presentation to actual customers. It could be as simple as establishing a process whereby all blog posts are reviewed prior to posting. It could be as complex as a Marketing Campaign with all that entails.

Let's talk about actual implementation and early results from experimenting with Quality Ownership at selected companies on selected projects.

Change is hard and this one is a radical departure from what makes your teams comfortable. Those of you who have had to shepherd a team through any type of radical methodology change will understand immediately what types of challenges you'll face when trying to implement Quality Ownership.

To help motivate you, there is a 95+ billion-dollar company out there that has done a great deal of interesting work in regards to what they refer to as Quality Assistance. They utilize their quality staff as mentors and teachers with responsibility for making the teams better. In other words, their quality staff do not produce anything – they teach others to provide test analysis and do the testing.

As most companies are not 95+ billion-dollar enterprises and cannot afford – or do not believe in – hiring staff who do not contribute any deliverables for a team, the Quality Ownership concept here includes the quality expert as part of the practical contributors to the team. The reason for that is purely due to human nature. Team members tend to disrespect and/or dislike those that merely "advise". A team member that produces a deliverable, just like every other member of the team, is viewed differently than a member that merely proffers advice and "doesn't get their hands dirty". It is also difficult enough to get staffing for development teams without trying to justify hiring for advisory roles.

It would be a disservice to you and to the field to not address the challenges and problems with implementing Quality Ownership. This is not a trivial change. It is highly likely everyone will hate it equally at first. Have any of you lived through pulling a team into agile processes for the first time? Yup. It's that bad.

The first problem has to do with Quality/Test Analysts and their view of their work. Many Quality Analysts identify and define themselves through actual physical testing activity. While having to perform the work under a constant cloud of pressure is not especially enjoyable, testing, in and of itself, is fun. Quality Analysts are accustomed to being the ones to find the errors, work overtime to save their team, and generally act as corporate heroes, even if just within their own frame of reference. That's hard to give up. What's more they don't trust anyone, especially developers, to actually execute their tests.

Quality Ownership shifts their focus to defining the tests and then potentially moving over to another team to do the same. In essence, it removes one of the most rewarding parts of their jobs. To add insult to injury from a Quality Analyst perspective, if AI ever steps up and starts to realize its potential for defining test sets for functional testing, the Quality Analyst should be using these tools to help get that task done. This is actually great for their careers and their futures. However, it also cuts into the most enjoyable part of their jobs.

I'm a "rip the band-aid off" kind of person, and our field needs to change. I think it best to Just Do It. In order to get the pain over with as quickly as possible, I believe the Quality Analyst should define the tests and then act as a "lead" for the testing itself, making test assignments, ensuring all of the tests were run, and that bugs were written and addressed appropriately. For the first few months, they shouldn't be permitted to do any physical testing.

This forces the rest of the team to step up do the testing and get out of the mindset where it's "QA's job". It's the team's job. Later, the Quality Analyst can also assign tests to themselves. But it shouldn't happen until the team has stepped up and accepted their responsibilities for testing their product.

What can genuinely be an advantage is that this solution allows for some techniques that are proven and historically efficient. It allows for session-based test management, which is immediate, effective, team-building, and fun for all involved. Think of it as a mob test. Everybody can sign into Zoom at the same time, running their tests, finding problems, talking to each other, and pounding out the tests really fast, with the Quality Analyst coordinating the entire event, answering any questions, and moving the session along. Much faster than one person trying to do it all themselves, isn't it?

If you are going to try Quality Ownership, it's worth mentioning (several times) that it needs unwavering, strong management support. No one gets a pass in terms of testing. Everyone is required to pitch in and do their part. One of the keys to success is moving from an "it is their job" to "it is our job" mentality. You can't do that if you allow someone to shirk their responsibilities.

On the development side, developers are going to kick back against this process. They are already tremendously busy and they definitely do not want to add testing to the mix. Particularly if that testing includes manual testing. That's actually a good thing, as part of the object of the methodology is to encourage test automation. Consider how much easier it has been for them to just hand off their work to someone else to test and move on to the next interesting thing to code. But the process does good things for the team. With everyone testing, the team finishes at the same time, rather than QA still working on the current sprint while everyone else moves on to the next.

There is a tendency for back-end developers to accept the methodology more easily than their front-end counterparts. This is likely because they are more accustomed to doing at least a portion of their own testing. Depending on the company and the skill sets of their quality staff, they may or may not be comfortable with "headless" testing – that is testing without a front-end feeding in data.

This type of testing requires use of a tool such as Postman or SOAP/SOAPUI. It's possible back-end developers in your company do their own testing.

Front-end developers, however, will have to become familiar with some sort of testing architecture and build their tests accordingly. This activity would be required for any type of automated or TDD/BDD methodologies, but it is human nature to resist change and busy staff members will push back even harder against learning or doing something new.   Again, this is going to require management to hang tough and not give any choices. Adapt or die. Your SDET personnel can help here both in setting up automation architecture, assisting the development staff with automation, or assuming responsibility for test automation, as best suits your company situation.

It is strongly recommended that when first implementing this process the Quality Analyst do substantial random checking to ensure all of the identified tests were executed and verified. It may seem laughable to some (a painful reality for many QA personnel) that a developer may not understand the need to test certain areas and will therefore opt not to run given tests. This is why developers are bad at defining functional test bases; they can't see the forest for the trees.

The companies who have experimented with this process confirmed they had this very scenario come up more than once and the random checks uncovered critical bugs. That means you need some process to ensure all identified tests are run and results evaluated. The good news is the problem goes away once everyone understands they must run every test identified and that they will be called out specifically if that doesn't happen – particularly if an error that should have been caught escapes into Production.

Again, management and executive management have to hang tough. It is very hard not to cave when your teams – or a particularly valuable member of one of those teams – tells you they can't handle testing. It is critical to insist new processes be followed. You, as a company, cannot allow individuals to dictate to you how to operate. Your decisions, no matter how difficult, have to be in the best interests of the company overall. Give your teams no choice but to join you in your experiment, make sure all of your management staff are on board and committed, and make sure the experiment lasts long enough that a few determined individuals don't torpedo your efforts.

What good things can you expect? The companies that have tried it found no increase in error rates in Production. Your development staff will become more user-savvy and thus, more useful to you and the team. The process allows for a graceful introduction of AI for unit testing, which means unit testing gets done and provides value to the team and less errors found later in the process. More tests get automated, again, a benefit to you and the company. All the team members finish the sprint at the same time. You don't need to hire more Quality Analysts every time you add applications or features. Depending on your set-up, you may or may not need SDET personnel for regular sprints, reserving them for setting up the automation architecture, creating an automated test base for smoke and regression tests, assisting a variety of teams, and the like. Estimation will improve to include testing efforts.

To conclude, I'd like to offer a quick step-by-step for implementation:

Communicate the process to the team and select one or more teams to spearhead the effort and allow you to fine-tune where necessary.

Get the QA/Test Analyst involved during design. The ticket must contain a structured outline/test plan that lists planned tests in order to be "ready to work". At the same time, utilize AI (if possible) to define unit tests. This can be done by either the Quality/Test Analyst or an SDET, depending on your company set-up. AI produces pretty test cases that are easier to automate.

The team acts as usual to pull tickets into a sprint.

Either the development staff or the SDET (again, depending on your company set-up) automates the unit tests as part of their development process and hooks them into your CI/CD process if you have one.

When all the pieces are done and ready for a final "functional" test, the Quality/Test analyst assigns tests to the entire team, arranges for test sessions, and runs those sessions. The team repeats as necessary, based on the errors found.

Several things can happen at this juncture – the sprint is "done". The functional tests can be expanded to hand over to SDETs, an offshore automation team, or stored as is for audit purposes. Expansion of a structured test outline or test plan is a good training tool for newbies, and it is appropriate work for an offshore team as well. As AI becomes better, it may be possible in the future to have the expansion task done by AI. It is considerably easier to do expansion after something moves to Production, as Production then acts as the "golden standard" – in other words, expected results are, in fact, expected results. It cuts down on questions, which equates to time, which equates to money.

When giving Quality Ownership a try, give it some time to truly assess your success or lack thereof. New processes take time to become second nature. Then, as with anything, really assess your results. What worked well? What didn't? Toss anything that didn't work, think about it, poke at it, change it, and continue to experiment.

Above all, let's all stop bitching about the same things year after year. We're better than that. Let's stop saying we do (X) when we clearly don't. We don't need to give lip service to stuff that doesn't work and has never really worked for us. We don't have to be rhinestones. We should all go out and change our quality world.

It's about time, don't you think?

www.ingramcontent.com/pod-product-compliance
Lightning Source LLC
Chambersburg PA
CBHW070505200326
41519CB00013B/2721